Charles T. Williams

Aero-Therapeutics

The treatment of lung diseases by climate - being the Lumleian lectures for 1893

delivered before the Royal College of Physicians, with an address on the high

altitudes of Colorado

Charles T. Williams

Aero-Therapeutics
The treatment of lung diseases by climate - being the Lumleian lectures for 1893 delivered before the Royal College of Physicians, with an address on the high altitudes of Colorado

ISBN/EAN: 9783337272463

Printed in Europe, USA, Canada, Australia, Japan

Cover: Foto ©berggeist007 / pixelio.de

More available books at **www.hansebooks.com**

AERO-THERAPEUTICS

OR

THE TREATMENT
OF LUNG DISEASES BY CLIMATE

BEING THE

LUMLEIAN LECTURES FOR 1893

DELIVERED BEFORE THE ROYAL COLLEGE OF PHYSICIANS

WITH AN ADDRESS ON THE

HIGH ALTITUDES OF COLORADO

BY

CHARLES THEODORE WILLIAMS,

M.A., M.D. OXON, F.R.C.P.

SENIOR PHYSICIAN TO THE HOSPITAL FOR CONSUMPTION AND DISEASES OF
THE CHEST, BROMPTON
LATE PRESIDENT OF THE ROYAL METEOROLOGICAL SOCIETY

London

MACMILLAN AND CO

AND NEW YORK

1894

RICHARD CLAY AND SONS, LIMITED,
LONDON AND BUNGAY.

PREFACE

I HAVE attempted in the following pages to sketch a scientific system of Aero-therapeutics, based on the combination of modern meteorology with clinical experience, each element of climate being duly considered in its bearing on Health and Disease.

Unfortunately our knowledge in these departments contains many gaps, which increasing experience may, it is to be hoped, gradually fill in.

Since delivery the lectures have been carefully revised, much additional material introduced, and an address on the climate of Colorado added.

2, UPPER BROOK STREET, W.
February, 1894.

CONTENTS

LECTURE III

BAROMETRIC PRESSURE IN ITS RELATION TO HEALTH AND DISEASE

THE HIGH ALTITUDES OF COLORADO AND THEIR CLIMATES

AERO-THERAPEUTICS

LECTURE I

BEFORE dealing with the subject of these Lectures—aero-therapeutics, or the healing influences of climate on mankind—it will be necessary to say a few words on climate generally and briefly to consider its principal factors. The chief factors of any climate are :

1. Latitude—naturally the greatest influence as describing the position of the sun towards the earth in a certain region, and thus determining the length and intensity of sunshine.

2. Altitude—by which the effects of latitude may be to some extent neutralised, for even in the tropics, at a height of 16,000 feet, snow and ice may exist; the temperature falling in ascending mountains 1° F. for every 300 feet.

3. The relative distribution of land and water, and especially the presence of vast tracts of either desert or ocean, the former accentuating extremes of temperature, and the latter tempering them.

4. The presence of ocean currents flowing from higher and lower latitudes (as the case may be) and qualifying thereby the climate.

5. Proximity of mountain ranges and their influence on the shelter from wind and on the rainfall.

6. The soil—its permeability or impermeability to moisture.

7. The rainfall—its amount and annual distribution.

8. The prevailing winds.

Such are the factors of a climate. We shall next come to its elements, which are five : temperature, hygrometry, atmospheric pressure, wind force, and atmospheric electricity.

The effect of altitude on temperature may be illustrated by the instance of South America, where in the tropics are to be found large cities enjoying a temper-

ate clime, owing to their altitude. Quito, the capital
of Ecuador, on the equator, at an altitude of 9,500 feet,
possesses the climate of perpetual spring, having a
mean temperature of 60° F. for every season; and
Santa Fé di Bogota, in New Granada, 8,648 feet,
about five degrees north of the equator, has a climate
resembling that of Malaga without the extremes. If
we want instances of the third factor of climate—the
influence of the relative distribution of land and water
—we find it best illustrated by a diagram which Mr.
Scott has modified from Dr. Supan's work (Diagram 1),
showing the equal annual range of temperature for the
globe; from which it appears that the range of temperature increases from the equator to the poles, and
from the coast towards the interior of a continent.
The regions of extreme range in the northern hemisphere coincide with the districts of lowest temperature
in winter, and, as might be expected, from the southern
hemisphere being largely covered with water, the
range is far greater in the northern hemisphere than
in the southern. The line of 20° F. range—the most
moderate, owing to the Pacific influence—extends up
the coast of Vancouver and California, crossing the
American continent in Florida and passing north-east
to the Faroe Isles, where it turns sharply, runs nearly
due south along the west of the United Kingdom and
Portugal to North Africa, where it reaches nearly to
the Gold Coast, from whence it skirts the north coasts
of the Indian Ocean to the south of China. Here is
prominently brought out the equalising influence of
the oceans, and specially of the Atlantic in its Gulf

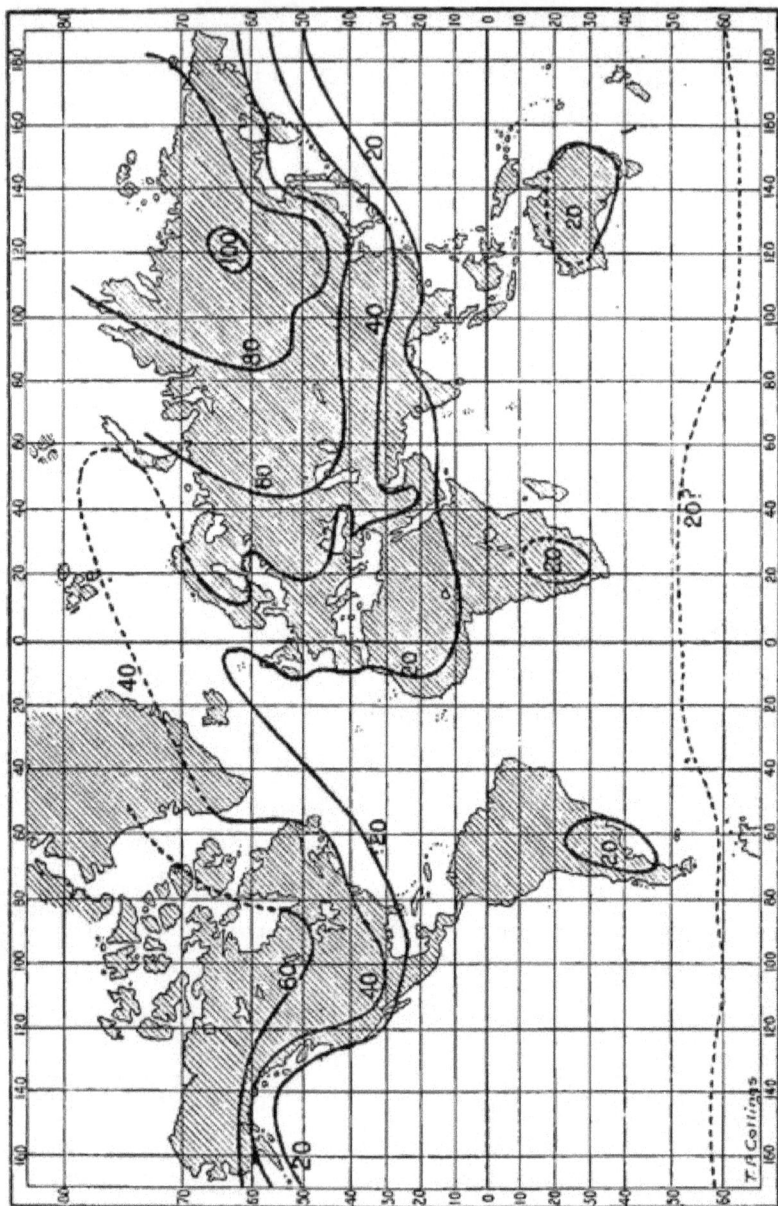

DIAGRAM I.—LINES OF EQUAL ANNUAL RANGE OF TEMPERATURE FOR THE GLOBE.

Stream, on the extremes of temperature, reducing what would be ranges of 40° or 50° to one of 20°.

On the other hand we see that on the purely land areas of Northern Asia ranges of 60° and 80° obtain, and in the region of Yakutsk in Siberia, 100° F. is reached, and at Werchojansk, Siberia, 120·4° F. Yakutsk itself has a temperature of 65·8° F. in its warmest month, and one of—44·9° F. in its coldest. In North America a mean range of 75 is reached in Northumberland Sound.

Inland climates tend to extremes, while those of coast and island are of a more or less temperate character. In the interior of continents the range in mountainous districts diminishes with the height above the sea,[1] and in the middle and higher latitudes of both hemispheres the western coasts have a less range than the eastern : the two exceptions to this rule being Greenland and Patagonia, which owe their special climate to the overwhelming cold of Arctic currents.

The fourth factor, ocean currents, is perhaps the most important of all, and certainly so to Great Britain and Ireland, for without the influence of ocean currents our climate would resemble that of Labrador.

The deep sea researches of the *Lightning*, *Porcupine*, and *Challenger* expeditions have established now beyond doubt that there exist both in the Atlantic and Pacific Oceans (1) a superficial layer of water extending to a depth of 500 to 600 fathoms, the temperature of which is regulated by surface currents arising from

[1] This is well shown in Pike's Peak, where the mean daily range is about half that noted on the prairie-plain at its foot.

periodic and variable or from permanent winds, and (2) a deep layer of far greater extent, which fills up the trough of the ocean, and does not vary greatly in temperature at different seasons, and is always below 39° F. This is a mass of cold water, which is constantly moving northwards from the Antarctic towards the Arctic pole, and is evidently an indraught from the Southern Ocean, of which both Atlantic and Pacific Oceans may be regarded as inlets or gulfs.

Thus far the deep layers of the oceans; let us now turn to the superficial currents, which may be divided into (1) warm equatorial currents, flowing from east to west, the result of the north-east and south-east trade winds, and (2) the polar currents flowing from the Arctic or Antarctic circles.

THE GULF STREAM

The most important equatorial current is the Gulf Stream, which is produced by the action of the north-east and south-east trade winds, and flows westwards to South America and, splitting in two off Cape St. Roque, part goes southwards to Cape Horn and the Falklands, and part flows north-eastwards along the coast of South America. This crosses the Caribbean Sea, and, making the circuit of the Gulf of Mexico, passes through the Straits of Florida, issuing as the Gulf Stream—a current 30 miles broad, 2,200 feet deep, with a temperature of 86° F. and a velocity of four miles an hour. It follows the American coast line, abuts on the cold Labrador current, than which it is warmer by 30° F., turns eastwards off Cape Cod in latitude 41°, and spreads, like a fan, over the Atlantic with diminished velocity. It divides off the Azores, part going southwards along the coast of Portugal and the Cape de Verde islands, joining the equatorial current, and the other portion skirts the coasts of Great Britain, Ireland, Scandinavia, and Spitzbergen.

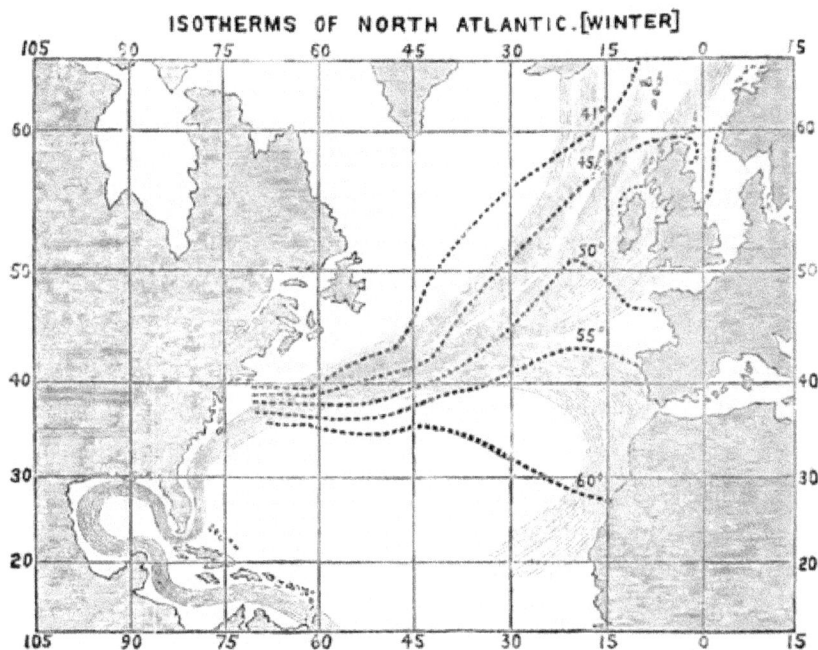

DIAGRAM 2.—GULF STREAM AND ATLANTIC ISOTHERMS.

The Atlantic steamers' winter course from New York to Queenstown lies for some time in the track of the Gulf Stream, and during a recent passage (in December, 1892) I noted the temperature of the water, which was tested twice daily. After the second day of the voyage it ranged from 56° to 58° F., the temperature of the air being generally 1° lower. The passengers used the water for bathing purposes without any artificial heating. In this track the weather was mild, and occasional showers fell. The Gulf Stream water has been found to move slowly off our west coast, not faster than 300 feet an hour, but its influence, especially in winter, though variously estimated, is undoubted, as every map of the North Atlantic isotherms will show; for the isotherm of $45\frac{1}{2}$° F. starts from the American coast at about latitude 38°, runs to the north of Scotland and far up into Norway to latitude 60°, causing a diversion of the temperature lines to the extent of 20° of latitude. According to the Rev. Dr. Haughton's estimate, the result of the Gulf Stream on climate in July is a cooling one, but in winter (January) it raises the temperature between latitudes 40° to 60° from 14·1° F. to 37° F. The effect on the shores of Great Britain in winter is best seen by a chart depicting the temperature of the waters of the British and Irish Channels during the winter months (say, in February), and showing how closely the air temperatures follow these.

It will be seen from Diagram 3, issued from the Meteorological Office, which is the result of numerous observations of the Channel water made at the various lightships and coastguard stations, that the temperature of the British and St. George's Channels varies according to the proximity of the Gulf Stream; the nearer to the stream, the warmer. The difference in water temperature between the sea off Scilly Isles and that off Sussex amounts to 5° F., the sea off the

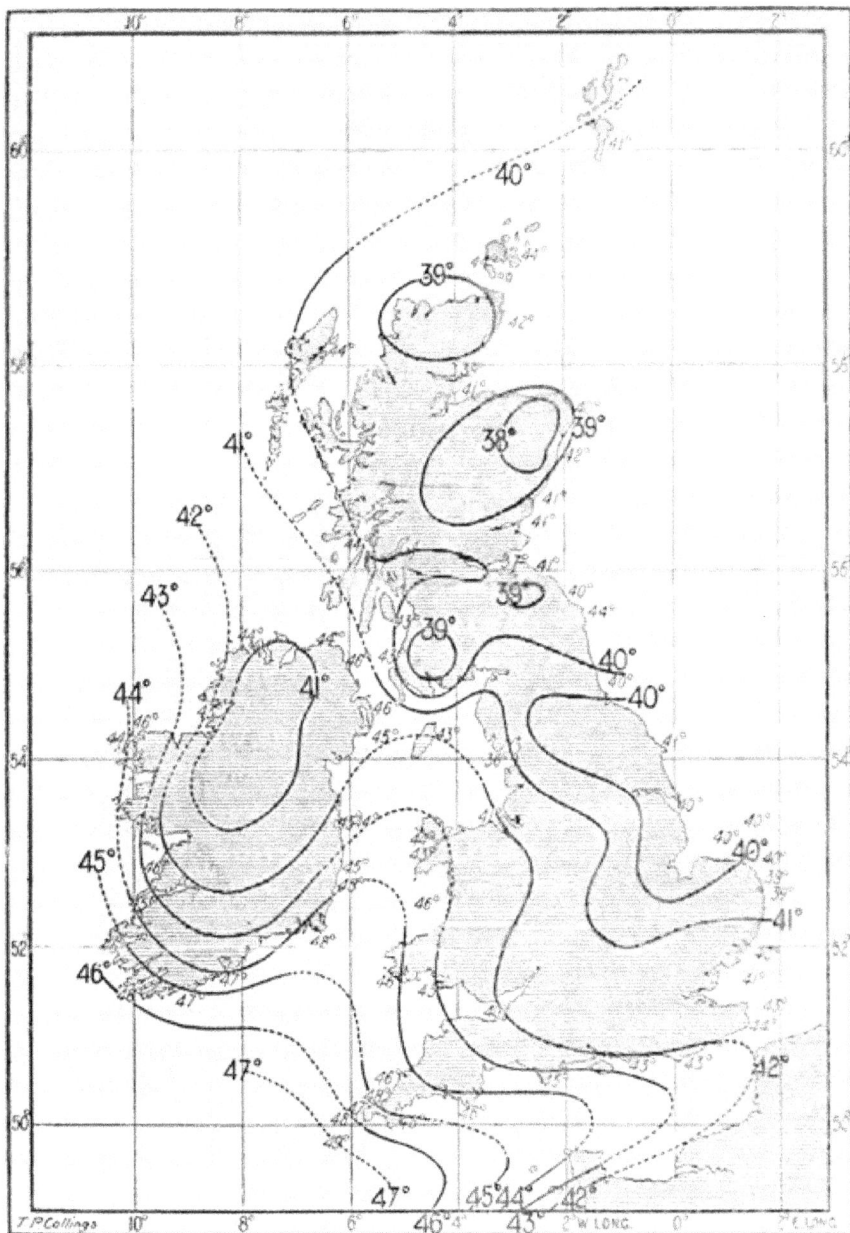

DIAGRAM 3.—MEAN TEMPERATURES OF AIR AND SEA FOR BRITISH ISLES IN
FEBRUARY.

Norfolk coast being still cooler, and showing a difference of 10° F. from that of Scilly. The air
temperatures follow the marine, being from 1° to 2° F.
lower. There are few better instances of the warming
and equalising effects of an ocean current than this,
by which a northern island is made to participate
to some extent in the warmth generated in the
tropics.

That the Gulf Stream influence extends to the
coasts of Great Britain, Ireland, and Scandinavia, and
even further, is proved by the North Atlantic
isotherms already given, by the warmth of the water
in the British and St. George's Channels, by the
drift-wood and tropical products found on the coast
even of Spitzbergen, and lastly by the temperature
soundings of Professor Mohn, of Christiania, off the
Trondhjem and Sogne Fiords, which prove the
existence of a warm surface current on the Norwegian
coast, ranging from 60·8° F. to 61° F.; and also that
it extends over a considerable area. Owing to the
North Atlantic being, to a great extent, a basin
closed to the northwards, the current of the Gulf
Stream, on reaching the barrier, is turned southwards
in a southern eddy, so that there is a certain tendency
for the hot water to accumulate in the northern basin,
and to bank up along the north-east coasts.

THE NORTHERN EQUATORIAL DRIFT IN THE PACIFIC

In the Pacific a similar phenomenon is observable.
The northern equatorial drift, produced by the north-

east trade, directs its course to the East Indian Archipelago and to the coast of New Guinea, and there divides, the southern part flowing on till it strikes the coast of Australia, and there turns eastward again, and the northern portion, known as the Kuro Siwo, or Great Black Stream of the Japanese seas, sweeps upwards in a north-easterly direction outside the chain of islands formed by the Philippines, the Loo Choo group, and Japan, and turning westwards towards the coast of North America, strikes it on the south side of the promontory of Alaska, and, flowing down the American coast, eventually joins the equatorial drift.

This warm stream, which is intensely salt, is estimated by Dr. Haughton as being nearly three times the size of the Gulf Stream, but never attaining the velocity of the latter, from its waters not being confined in a narrow channel. A large portion of it reaches the American coast, and produces, according to Von Baer, such an effect on its climate, that on the southern side of the narrow promontory of Alaska, humming birds are found, while the northern shores, which are washed by the cold current from Behring's Straits, are visited by walrus.[1] This warm current bends the isotherms northwards, as is seen in Diagram 4, and gives Sitka and the coasts of British Columbia their immunity from the ice which imprisons the harbours of Asia in corresponding latitudes.

To this current the fine climate of South California owes its equability, and its protection from the great

[1] Scott. *Elementary Meteorology*, p. 306.

extremes of its inland neighbour, Arizona. The Kuro
Siwo current, like the Gulf Stream, has a cooling
influence in summer, and the effect of the westerly
winds it produces may be seen in the contrast between
the July temperatures of the Pacific Coast stations
and of those at some little distance inland. For

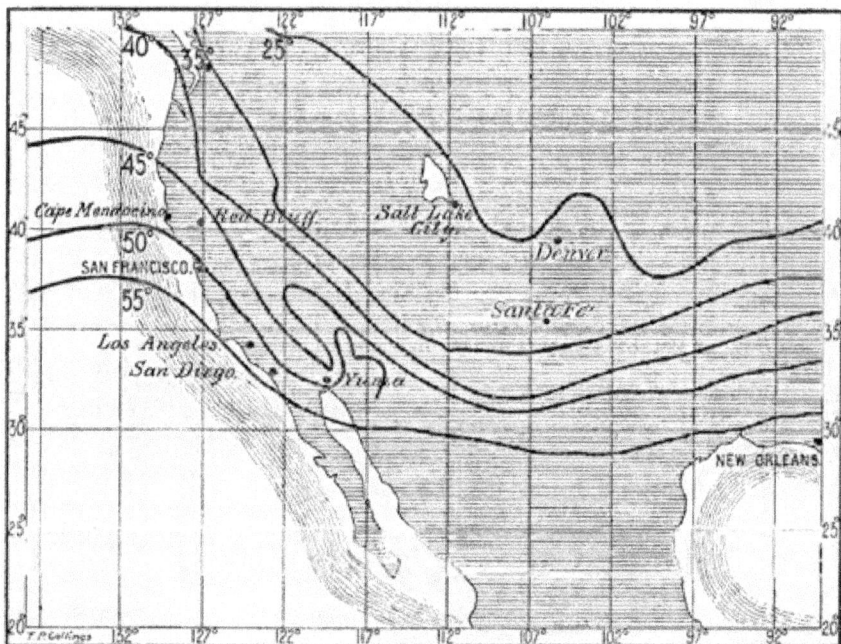

DIAGRAM 4.—INFLUENCE OF KURO SIWO CURRENT ON PACIFIC COAST
WINTER TEMPERATURES.

instance, at San Diego, in South California, the July
mean is 67° F., whereas at Yuma, in Arizona, at the
same latitude, and less than 200 miles inland, it is
92° F., a difference of 25°; and again at Cape
Mendocino, on the Pacific Coast, compared with Red
Bluff, less than 100 miles inland, there is a difference

of 28°. On the other hand, in January, the Pacific
Coast temperatures are higher than the inland ones,
that of San Diego being 54° against 50° of Yuma,
and that of Cape Mendocino being higher than that
of Red Bluff.

THE INDIAN EQUATORIAL CURRENT

In the Indian Ocean the equatorial currents are
to some extent, embayed, as there is no exit on the
northern side, and they become drift currents depending
on the monsoons and changing with them, but the
main equatorial current, east of the line, splits on the
coast of Madagascar, and the larger portion of it
flows down the Mozambique Channel, and becomes
the warm Agulhas current which washes the east
coast of the Cape of Good Hope.

COLD CURRENTS

The principal cold currents are two, flowing into the
Atlantic from the north circumpolar sea, according to
Sir George Nares, one passing along the west coast of
Baffin's Bay, and the other along the east coast of
Greenland at the rate of four miles a day. The one
from Baffin's Bay flows along the American coast, and
is to be distinguished as a surface current as far south
as Cape Cod. For part of its course it abuts on the
Gulf Stream. The American Survey shows that it
then dips under, and is to be found, at a depth of 100
fathoms, flowing under the Gulf Stream in an opposite

direction, and dividing in the Strait of Florida into two streams, one portion passing under the hot Gulf Stream into the Gulf of Mexico, and the remainder coursing round Cuba. This current, flowing southwards, hugs the American coast, and influences the temperature as far south as Cape Cod in latitude 42°, contributing to the cold climate of Labrador, Nova Scotia, and Newfoundland, and Eastern Canada.

The cold currents from the Antarctic Pole are (1) Humboldt's current, which appears on the western coast of South America, and lowers the climate of Chili and Peru, and another Antarctic current strikes on the West African coast near Cape Town, making the temperature about 20° lower than at the corresponding latitude on the east coast, and filling Table Bay with a profusion of fish, most of the edible varieties inhabiting cold water.[1] Both these currents exercise a marked influence on the adjacent coasts. The above description of ocean currents must be taken with the proviso that the whole subject is undergoing fresh investigation, and it is not impossible that our knowledge will be considerably extended by such investigation.

MOUNTAIN RANGES

The proximity of mountain ranges and their influence on the rainfall and on the shelter from winds is the fifth of our factors, of which plenty of examples are at hand. The neighbourhood of mountains, and especially the environment of lofty ranges, has been

[1] Scott, *op. cit.*

ascertained to increase the rainfall, except under
certain conditions of protection ; for, if the range rises
abruptly to a great elevation and the locality be to the
lee-side of it, the air currents prevailing to the wind-
ward are forced upwards, and, becoming lowered in
temperature, are obliged to part with their moisture,
and thus pass over the range as dry winds. This is
the case with the Colorado sanitaria, which are under
the lee of the Rocky Mountains and possess exceedingly dry climates.

Many sanitaria owe their exceptional climate to the
position of mountain ranges to the north of them ;
others have mountains to the east and west as well,
and enjoy remarkably calm atmospheres, but it must be
borne in mind that the greater the protection from
wind the less is the period of exposure to sunshine,
and in many cases the shorter the hours of daylight.
In some of the Swiss sanitaria the sun rises in
December at 8.45 A.M and sets at 2.30 P.M.

The influence of mountain ranges is also shown in
difference in day temperature of their southern slopes,
and of the valleys below. The mountain slope is
warmer, being more open to the sun's rays, and the cold
air, being heavier, collects at the bottom of the valleys,
which are consequently colder.

SOIL AND VEGETATION

The sixth factor is soil and vegetation—a purely
local one ; but their effect on climate is sometimes
surprising. The influence of soil on the accumulation

of moisture is well known, and the close connection of non-permeability of soil with the production of diseases, such as phthisis, has been proved by our distinguished Fellow Sir George Buchanan, by Dr. Bowditch, and others. The effect of different soils on the sun's rays, and their relative conducting power, is also of great interest. It would appear that light loose soils, such as sand or gravel, from their particles not being closely packed together, imprison large quantities of air in the interstices, and thus reduce the heat-conducting power of the soil ; whereas heavy, dense soils, such as clay, from their particles being packed together, are better conductors of heat. Therefore light loose soils are subject to high temperatures and to a greater degree of frost near the surface than dense, heavy soils ; but, on the other hand, heat and cold in the form of great frosts and extra temperatures do not penetrate so far down into light as into heavy soils. According to Dr. Buchan,[1] some experiments made in Scotland showed that at three inches below the surface the temperature fell to 26·5° F. in loose sandy soils, and at a depth of twelve inches the freezing point was only once reached. On the other hand, in clay soils at a depth of three inches, the lowest was 28° F., and at twelve inches the temperature often fell to freezing, and even at twenty-two inches 32° F. was once recorded.

In the same way the solar heat does not penetrate so deeply into sand, which is a bad conductor, and instead of conducting the heat to a greater depth as

[1] *Handybook of Meteorology.*

rock, and loam, and clay soils do, it accumulates it, and hence the great heat of the soil of the desert, which has been known to rise to 120° F., 140° F., and even to 200° F. When these particles of sand are lifted into the air by winds, as during the prevalence of the terrible Simoon, the atmosphere has been known to rise to 125° F. in the shade, and thus the deserts of Asia and Africa are stated to have a mean summer temperature ranging between 92° and 95° F. At night, however, the soil gives rise to marked radiation and consequent great lowering of temperature, and in this way we get immense thermic ranges. The covering of the soil with vegetation protects it from the sun's direct rays, and the temperature of plants exposed to the sun does not rise so high as the soil itself, because much of the heat is lost through the large evaporation which takes place from the leaves and stems, and which gives rise to air currents tending to reduce the temperature. The result is that the heat is more evenly distributed over the twenty-four hours, and is less intense in the hottest time of the day.

The effects of forests on the temperature and rainfall have often been discussed, and the general conclusion arrived at is that by retaining and absorbing the moisture, they moderate heat. According to Buchan, trees acquire their maximum temperature at 9 P.M., instead of between 2 and 3 P.M., the maximum period of air, and then they radiate it slowly at night.

With regard to the influence of forests on increasing rainfall, proof of this has been shown, first by the increase of rainfall in a district following extensive tree planting, and the reverse, namely, denudation of a region of trees, on its ceasing to be cultivated, being followed by reduction of moisture, as the drying up of rivers ; for example, that of the Euphrates and Scamander. A good instance of the influence of increased forest growth on climate is this one taken from Dr. Buchan's able work : " The valley of Araqua in Venezuela is shut in on all sides, and the rivers which water it, having no outlet to the sea, unite and form Lake Tacariqua. This lake, during the last thirty years of the last century, showed a gradual drying up, for which no cause could be assigned. In the beginning of the present century the valley became the theatre of deadly feuds during the war of independence, which lasted twenty-two years. During that time land remained uncultivated, and forests, which grow so rapidly in the tropics, soon covered a great part of the country. In 1822 Boussingault observed that the waters of the lake had risen, and that much land formerly cultivated was at that time under water.

From observations at the meteorological stations scattered about in the Great Lüneburg heath, North Germany, a tract of country which has been gradually re-forested at the rate of 1,000 to 1,500 acres a year, it appears that, while at first the rainfall at Lintzel, one of the forest stations, was only 80 per cent., of the stations outside the forest conditions, at the end of seven years it rose (as seen below) from a deficiency of 20 per cent. to an excess of nearly 4 per cent.

1882.	1883.	1884.	1885.	1886.	1887.	1888.
81·3	86·3	95·2	99·8	100·6	103·7	103·9

THE RAINFALL.

This is a factor which unfortunately can make itself very unpleasantly felt, and has been largely studied of late by Dr. Hann, Mr. Symons, and others, and on the whole the causes of excessive and deficient rainfall have been fairly explained.

According to Mr. Scott, the three great agencies in the precipitation of rain are :

(1) Ascending currents which, being lowered in temperature by their ascent, are compelled to deposit their moisture.

(2) The contact of warm air with the cold surface of the ground.

(3) The mixture of masses of air of different temperatures.

Of the first, wind coming across the ocean and striking a lofty range, which forces it upwards and causes it to deposit its moisture, is a good instance, such as may be seen in the south-west monsoon, striking the Khasia Hills, in Assam, and producing the tremendous rainfall of 493 inches at Cherrapunji ; of the second, the contact of the warm south-west wind with our own Cornish and Devonshire coasts, which being much colder in winter, causes rain precipitation. The third agency is to be seen in the regions where ocean currents of varying temperature meet ; the aërial currents accompanying them commingle and cause heavy deposition of moisture, as may be seen in the

fogs off the Newfoundland banks, where the Gulf
Stream and Baffin's Bay currents meet.

When the wind is the rain-bringer, as it usually is,
localities to the lee of the mountain range have small
rainfalls, the moisture having been deposited in the
mountains.   In regions surrounded by mountain chains,
such as Utah, in the United States of America, and
Gobi, in Siberia, the dryness produces a desert, the
mountains having drained the winds of all moisture
before reaching these tracts.

The rule about rainfall is, that other conditions
being equal, it decreases in quantity from the equator
to the poles, while the number of rainy days increases.
Tropical countries have rainy seasons, when rain falls
in large quantities for weeks, the rest of the year being
free ; and in temperate countries rain falls all the year
round.   In this country the rainfall, according to Mr.
Scott, depends " on the Atlantic influence, and on the
somewhat irregular succession of barometric depressions
and anticyclones which are constantly moving over
the earth's surface in the temperate zone."   The
rainiest month in this country on the west coast is
January, and on the east coast and London it is
October ; the driest month being March.   In Europe,
north of the Alps, more rain falls in summer ;
south of the Alps, more falls in autumn.   On the
Riviera, for instance, there is hardly any rain in
summer, and nearly half of the annual amount falls
in September, October, and November.   As we pass
eastward, the summer rainfall increases.   In north-
west France it is 24 per cent. ; in northern Prussia,

36 per cent. ; in central Russia, 38 ; and in the Ural mountainous district, 53 per cent., more than half the total amount falling in summer.

Mountains exercise considerable influence in condensing moisture, and causing rain precipitation, and this property appears to increase with the elevation up to 3,000 feet or 4,000 feet in Europe, and then to diminish above this level.

WIND

The last, but by no means the least, climatic factor is wind, and the prevailing wind is often the key to the climate of a locality. Winds arise from differences of atmospheric pressure, the wind blowing from a region of higher pressure to one of lower pressure, this continuing until equilibrium is established, and it would appear that the majority of the winds which prevail over the globe may be accounted for by the appearance of certain spots of barometric depression during certain seasons of the year. In the month of January the pressure is low in the neighbourhood of Iceland, but rises in a south-westerly direction on approaching the American coast. It also rises to the south over the Atlantic, and to the east over Europe and Asia.

Now, observations show that this remarkable depression of the barometer determines the prevailing winds over a large and important part of the earth's surface, for during the month of January in North America there is a decided preponderance of winds from the N.W. in the Atlantic between Great Britain and America, as well as over France, Belgium, and the south of England. The direction is nearly S.W. at Dublin, and in the south of Scotland it is W.S.W., and at Copenhagen it is S.S.W. At St. Petersburg it is nearly S., and at Hammerfest near the North Cape (Norway) it is S.S.E. A comparison of these directions of the winds with the iso-barometric chart for January explains matters, and as Mr. Buchan[1] puts it, "All the prevailing winds in January over this extensive region are the simple expression of the difference of barometric pressure, which prevails in different parts of the region. The whole of the atmosphere flows in towards and upon the region of low pressure round Iceland, not directly towards the region of lowest pressure, but in a direction a little to the right of it."

In the interior of Asia during the same period, high barometric pressure prevails, which gradually diminishes on all sides on approaching the coast, and consequently the winds appear to flow *out* of this region of high pressure. For example, at Calcutta the prevailing winds in winter are N., at Hong Kong E.N.E., at Pekin N.W., at Hakodate in Japan W.N.W., and at Bogostovsk, (Ural Mountains) S.W. We have various kinds of winds : (1) permanent, such as the north-east and south-east trades, which blow towards the equator from the poles, to replace the ascending heated air of the tropics, and owing their easterly direction to the earth's rotation. These prevail continuously, but shift their area of prevalence

[1] *Handybook of Meteorology.*

during different seasons of the year; then come the
seasonal or periodic winds, of which the north-east
and south-west monsoons are a good example. These,
according to Sir Joseph Fayrer,[1] arise in April by the
heated air over the continent of India being replaced
by comparatively cool currents laden with moisture
coming from the Indian Ocean from Africa to Malacca.
This is the south-west monsoon which, rising to higher
regions on being intercepted by the mountain ranges,
condenses its moisture in rain on the Western Ghâts
and on the Coast of Aracan. Following a north-east
course, it loses its influence and its moisture as it
approaches the northern limit of the continent, and
about October there follows a reversal of the current,
which then blows southward as a dry current, till on
the Coromandel coast it brings moisture from the
Bay of Bengal, which falls as rain on the coast of the
Carnatic and Eastern Ghâts, while some parts of
India receive rain with each monsoon.

Other winds are variable, such as most of the winds
of the temperate regions, but in some climate special
winds are quite characteristic, such as the mistral of
the south of France, the sirocco of southern Italy and
Sicily, the khamsin of Egypt, the harmattan of the
Sahara Desert, and the south-west and east winds of
this climate.

But we may be quite certain that beyond the use
of winds for propelling vessels and machinery, they
serve a distinctly hygienic object in dispersing
noxious exhalations, whether animal or vegetable, in

[1] *Rainfall in India.*

permitting free evaporation, and thus preventing accumulation of moisture, and in maintaining the circulation of air which is necessary for the purification of the atmosphere.

TEMPERATURE

We now come to the elements of climate, which have been enumerated above, and let us first consider temperature in relation to man's well-being. Experience has shown that the natives of temperate countries, such as our own, can endure great extremes of heat with only a small rise of body temperature, provided the atmosphere be dry, the skin acting freely, and the period of probation not too long. Tillet states that at the town of La Rochefoucault at the bakeries the female assistants, as a rule, remained ten minutes in the oven at 132° C. (301·6° F.) without much suffering. Messrs. Blagden, Fordyce, Banks, and Solander bore a temperature of 260° F. with the small rise of 2½° F. as long as the air was dry and perspiration free, but if the air became moist and evaporation was hindered, the temperature of the body rose 8° F. In the same oven with the observers, eggs became hard in twenty minutes, a beefsteak was cooked in half an hour, and water boiled.

The effect of excessive sun heat has been sometimes very disastrous to troops on the march, as was seen during General Bugeaud's expedition in the province of Oran, Algeria, in 1836, when the sun-temperature rose to 161° F., and in a few hours eleven soldiers committed suicide and 200 men were attacked with congestion of the brain.[1]

Sunstroke or insolatio is generally the result of solar or artificial heat in tropical, but occasionally in temperate, climates, and, according to Sir Joseph Fayrer, the most frequent cases are those coming on in houses, barracks, and tents, away from the solar rays ; and the subjects most likely to be attacked are those debilitated by disordered health, by dissipation and over-fatigue, rather than those of vigorous constitution or those who have undergone acclimatisation.

He enumerates three varieties of sunstroke : the first showing itself in exhaustion and syncope ; the second being a condition of shock in which the nerve-centres, and especially the respiratory, are affected, causing rapid failure of the respiration and circulation ; and the third, of which the main feature is intense pyrexia due to vasomotor paralysis and to the nerve-centres being over-stimulated, and then exhausted by the action of heat on the body afterwards. Recovery from the first is common, but the second form is far more serious and often fatal, and in the third form, where the temperature rises to 108° F. or 110° F., the results are often fatal. The disease may be caused by the heat of over-crowding or bad ventilation in tropical countries, whereas the second form is generally due to the direct action of the sun's rays on the head and spine. Fayrer[2] remarks : " Hindu natives, on their bare heads and necks, endure an amount of sunshine which

---

[1] Boudin, *Géographie Médicale*, vol. i. p. 397.

[2] Quain's *Dictionary of Medicine:* " Sunstroke."

would be fatal to a European; but if the temperature rise above a certain standard all succumb, the natives suffering like others, and dying of sunstroke.

The atmosphere of the plains of India, and especially of Bengal, contains a large amount of moisture, which makes the endurance of heat more difficult, but in the north of India and the dry regions of the United States to the west of the Rocky Mountains great degrees of heat, that is, 118° to 128° F., are tolerated from the dryness of the atmosphere, and, as General Greely, the late Chief Signal Service Officer of the United States Army, says,[1] "the inhabitants of the Atlantic and central stations of the country hear with amazement of the extremes of heat reported from the arid regions of Arizona and South Colorado as being within the bounds of human endurance, and cannot believe that the ordinary avocations of life can be pursued without inconvenience; however, the explanation lies in the climate being cloudless and dry, and promoting rapid evaporation, and consequently no suffering ensues, and sunstrokes are unknown."

Sunstroke is almost unknown on board ship, even in the tropics, provided the vessel is in mid-ocean, though the sun's rays are very powerful.

The effect of great heat on the lungs is to reduce the number of respirations, according to Rattray, from 16·5 in temperate regions to 13·74, and even to 12·74 in the tropics, accompanied by a slight spirometric increase, not enough to account for the decreased number of respirations, and so the respiratory function

[1] *Report*, 1890.

is reduced at least 8·43 per cent. As the late Professor Parkes puts it, " if 10 ounces of carbon are expired in the temperate zone, only 8·17 ounces would be expired in the tropics." The water exhaled by the lungs is also diminished, and the observations of Parkes and Francis show that the lungs of Europeans dying in India are lighter after death than the European standard. This might be expected from the diminished use of these organs, as witnessed by the lessened number of inspirations. The heart's action is not perceptibly quickened in the tropics, and the pulse is not faster than in temperate regions. The digestive powers are weakened, appetite fails, the liver becomes congested and undergoes changes which may end either in induration or abscess. The urine is diminished in amount, the urea is reduced, possibly from the small amount of animal food consumed. The nervous system is depressed, especially if humidity is combined with great heat, but perspiration is abundant, and has been estimated to increase 24 per cent. in the tropics.

Protracted residence in the tropics appears to exercise a depressive influence in lessening nervous energy and impairing the functions of digestion, assimilation, respiration, and blood making, and the power of forming new and healthy tissue. The tint of the skin and the conjunctivæ, and an appearance of premature age in Europeans long resident in the tropics, all go to confirm this conclusion.

The human body also seems capable of enduring great cold, when proper precautions as to food and clothing are adopted, and the atmosphere is still ; any wind renders even moderate degrees of cold unendurable. In the Arctic regions Captain Parry noted the thermometer as low as − 55° F., or 87° below the freezing point ; and Sir George Back at −70° F., or 102° below freezing point.

During my visits to the Engadine in winter I have often exposed myself with impunity to a temperature of −4° F., or 36° below the freezing point, and have noted the fact of delicate invalids sleeping with their windows open in that temperature, even when it is as low as −11° F. (43° below freezing point) and yet suffering no harm. The lowest temperature recorded [1] is −90° F. at Werchojansk, in Siberia, latitude 67·5° N. ; the average temperature for the month of January, 1885, being −63·9° F., and for February −84·3°, for March −77·4°, and for December −78·2° ; the maximum of December 33° F. This cold station lies in the valley of the Iana, 330 feet above sea level, and, owing to its latitude, the sun is absent altogether during December, while its elevation above the horizon for the rest of the winter is so slight that the effect of the direct sun's rays are unable to counteract the intense cold caused by the radiation.

In North America the records are also very low, reaching at Poplar River, Montana, to −63·1° F. in January, 1885, besides embracing a large number of records at various stations from −40° to −54°, and many of these interior stations give lower records than the Arctic ones.

The question arises—how do these low temperatures

[1] Greely's *American Weather*, p. 121.

affect the human body? If exposure to cold be prolonged, and the circulation and thermogenic powers cannot be maintained, the blood vessels, especially the smaller arteries and capillaries, become contracted, and no longer permit the passage of blood corpuscles, and all physiological and chemical changes are arrested. Various parts, especially the extremities, become starved, and hence death of these parts takes place by frost-bite or gangrene, appearing first in the fingers and toes. The effect of cold on the blood has been well demonstrated by Drs. Bristowe and Copeman in a case of paroxysmal hæmoglobinuria, communicated to the Medical Society in 1889,[1] where careful determination of the number of red corpuscles before and after the application of cold to the patient, showed that exposure to cold air produced a temporary reduction in the number of red corpuscles of 129,000 to 824,000 per cubic millimetre. The same effect was produced by a cold bath, and even by plunging the hands in cold water, the blood-making process in each case being gradually restored on returning to a warm atmosphere. Prolonged exposure to extreme cold gives rise to languor, lowering of the sensibility, and the individual loses all power of reaction, and sinks to sleep—often to wake no more—as is witnessed on long marches through the snow, the form of death being coma. Another result of the cold is to produce brain excitement, delirium, incoherency and thickness of speech, the symptoms resembling those of intoxication. Death occurs generally by syncope or asphyxia.

The capability of man to endure variations and
extremes of temperature has been proved to be very
great, for General Greely states that at Fort Conger,
U.S.A., in February, 1882, he experienced the low
temperature of −66·2° F., and, at another time in the
Maricopa Desert, Arizona, he saw noted the air
temperature of 114° F., while the metal of his aneroid
beside him as he rode assumed a temperature of
144° F.[1]

Having sketched out one of the elements of
climate—temperature—we must consider it in reference
to its influence on lung diseases, premising that the
beneficial effects of a pure atmosphere are not to be
assigned to one kind of climate only, as it is well
ascertained that patients, and specially consumptive
patients, have recovered in all climates—hot and cold,
dry and moist, clear and foggy—and that the arrest
of tuberculosis may be due to the patient's improved
constitutional powers, fostered by more favourable
surroundings, of which a propitious climate is one ;
but where we find a large number of similar cases
distinctly improving under change of climate, and the
accompanying increased facility for outdoor life and
exercise, and where we find also the percentage of improvement larger than among similar cases under
different climatic conditions, in the absence of other
factors, we may fairly assign the improvement to the
change of climate.

Before proceeding further, it may be well to allude
to the question of immunity from disease, and specially

[1] *American Weather*, p. 121.

from consumption, with which some climates have been credited, and which has been proposed as a basis of climate selection. The localities stated to be immune vary so greatly in climate conditions, some being of high altitude, some below sea-level, some with tropical heat, and some of intense cold, that it is impossible to discover any common qualities possessed by them. This is a subject which I discussed at length in the Lettsomian Lectures[1] before the Medical Society of London some years ago, and I may refer the Fellows to the arguments then adduced ; but the great point to bear in mind with regard to immunity from phthisis of any given place, is whether the conditions of life there are such as to foster or produce consumption, and, if so, does this locality remain immune, because this test alone would be a reliable argument for its recommendation on the ground of immunity.

We will now proceed to deal with warm climates but it will be of no service to discuss absolutely tropical climates, as experience teaches that these are not desirable for the treatment of disease : but it may be of considerable use to state what has been effected by moderately warm climates on patients specially selected for such treatment.

Warm climates may be divided into warm moist and warm dry.

[1] *Influence of Climate in Pulmonary Consumption.*

WARM MOIST CLIMATES

The best type of the warm moist climates is Madeira, as it enjoys the advantages of a marine atmosphere : the air is therefore permeated with a certain amount of saline vapour. Sixty years ago Madeira was the *beau idéal* of a climate for consumption and lung disease, and enjoyed perhaps a higher reputation than any other sanatorium. The annual mean temperature is a little above what we heat our houses to in winter. The winters are warm and the summers cool, the difference between winter and summer mean temperatures not exceeding 9° F. There are no cold winds, and only an occasional *leste* or hot wind from the desert. The nocturnal radiation is slight. The relative humidity percentage is large and the rainy days numerous, equability being the great feature of the climate. The principle of sending patients to Madeira was to keep them in an equable atmosphere, in a sort of aërial warm bath, which soothed the respiratory passages (and how acceptable warm moist air is to irritable lungs!) and promoted expectoration, permitting also of much sitting and lying out of doors. Unfortunately this soft atmosphere had often an injurious effect on the general health, inducing langour, loss of appetite, and even diarrhœa, and apparently promoting progress of the tuberculous disease.

My statistics published in the *Medico-Chirurgical Transactions*, vol. lv., show that among sixty-three consumptive patients who spent one or more winters on the island, 53·01 per cent. improved, 14·28 per cent. remained stationary, and 31·91 deteriorated, and this unfavourable result was arrived at, although 63 per cent. had tuberculosis without excavation, a favourable outlook, and in not more than 40 per cent. was the disease bilateral. However, many of the improved class not only improved, but improved greatly, and several of the excavation cases also, which formed nearly one-third of the whole number, showed signs of contracting cavity, proving that undoubtedly Madeira does suit some phthisical patients. Dr. Lund's statistics of 100 phthisical patients are somewhat more favourable, but Dr. Renton's are less so than my own. It will be remembered that of twenty phthisical patients sent by the Brompton Hospital to Madeira for one winter, only three improved, one died, and the rest returned to England worse than they started, and yet these were cases carefully selected by the medical staff as most likely to benefit by the climate.

For the majority of consumptives this sort of climate does more harm than good, but for the catarrhal form of phthisis it is, as my statistics show, a distinct success.

Chronic bronchitis with emphysema, bronchial catarrh, pulmonary congestion in elderly persons, unconnected with heart disease, are wonderfully relieved, and the patients often quite lose their symptoms. The influence is most marked in chronic bronchitis, the expectoration becomes easier and at first more abundant, then gradually diminishes, and the cough becomes less troublesome and in due time ceases. Sleep is sounder,

and the patient rejoices in being able to breathe more
easily.   Bronchial asthma often does well in Madeira,
especially if associated with much catarrh.   I have had
three patients with spasmodic asthma, accompanied
with catarrh, who each passed one or more winters
without attacks there, and were able to take exercise
all the season.   My general impression of Madeira
results is that the patients who improved were those in
whom catarrh formed the prominent and most troublesome symptom of the case, whether the illness were
asthma, phthisis, or bronchial inflammation ; and also
that those patients who were able to ride improved
more than those who were always carried in hammocks.

THE CANARY ISLANDS.

The Canary Islands, in latitude 27° to 29°, including Teneriffe and Grand Canary, enjoy a climate
similar to that of Madeira, but somewhat warmer and
drier, and are much frequented by English pulmonary
invalids.   The winter mean temperature is 64·7°, the
coldest month (February) having a mean temperature of
62°F., and the warmest summer month (July) one of
76°, the total range being only 11° to 14°, and the rain
fall 13 to 15 inches, distributed over about fifty days.

This remarkable freedom from extremes of heat and
cold, Teneriffe owes to the presence of a layer of cloud
which forms round the Peak and screens the lower
portions of the island from the full power of the sun
at midday.   Dr. G. V. Perez[1] notes that with the

[1] *Orotava as a Health Resort.*

first breath of the trade breezes in the morning, small clouds form about the mountains and coalesce in a vapoury parasol, which, about 11 A.M., covers the upper parts of the island, to be blown away in the afternoon by the land breeze (*terral*), which prevails from 5 P.M. till 10 A.M. on the following day, and from its descent into lower levels becomes warm. This cloud-parasol, though it acts so well as a shade, is not so very dense, commencing at 3,000 feet and terminating at 5,000 feet, and is easily traversed by climbers of the Peak, who see it below them stretched out like a stormy sea, covering a great portion of the island. Above the cloud-layer, the climate, according to Dr. Marcet,[1] who carried out an interesting series of observations on the meteorology of the Peak in July, 1878, is a totally different one, and presents great extremes, rising to 83° F. in the day and falling to 28° F. at night, even in July. The atmosphere was exceedingly dry, and the difference between the wet and dry bulbs on one occasion amounted to 30·5° F. The wind is generally N.E. in the island, but at an altitude of 10,700 feet, westerly winds with a southern veer were noted, indicating the presence of the return trade current.

Mr. Ernest Hart,[2] whose charming description of the valley of Orotava and powerful testimony to the wonderful qualities of climate, attracted numbers of patients to the Fortunate Isles, writes : "From one year's end to the other, the variation of temperature

<hr />

[1] *Southern and Swiss Health Resorts*, p. 301.

[2] "A Winter Trip to the Fortunate Islands," *British Medical Journal*, April and May, 1887.

does not exceed 18° F., and this within the limits most
favourable to life. That is the whole magic of the
climate. There is no excessive heat in summer, no
cold in winter. Very small rainfall and that chiefly at
night. No chill at sunset, no heavy dews, no frost,
no sirocco. It is a climate full of geniality, with
neither bite nor burn. It has the charms of the
temperate zones without their fluctuations and their
drawbacks; the delights of southern continents without
their pests, such as mosquitoes, venomous beasts and
insects, their excessive heats, their miasmas, or their
heavy rainfall."

The chief features of the climate are : (1) its remarkable equability by day and night, in summer and in
winter, due partly to its insular position and partly to
the tempering influence of the cloud parasol; (2) its
small rainfall and small relative humidity percentage,
which only amounts to 59·9 per cent. It seems well
suited to patients who are liable to catch cold and to
those of feeble circulation.

The great advantage which Teneriffe offers is the
variety of sites for residence at different altitudes on
this mountainous island, rising as it does from the sea-
level to the height of 12,200 feet in the famous Peak,
and with a choice of villages and hotels at different
elevations it is possible to pass the whole year in the
island without suffering from the heat. According to
Dr. Perez, it were well if more accommodation could
be supplied in the region of the Cañadas, about 7,000
feet above the sea-level, which would furnish an excellent site for a high altitude sanitarium.

My experience of this warm climate is limited, but I have seen great benefit derived by patients suffering from asthma and chronic bronchitis, but the few cases of phthisis which I have sent there have not prospered. Warmth and equability combined with dryness appear to be the chief attractions, but the bracing element seems to be wanting.

EGYPT

Of dry warm climates the most typical is that of the deserts, such as are to be found in the centre of Australia, that of Gobi in Chinese Tartary, the great deserts in the United States, and the tract which stretches from the great Sahara through Arabia into Persia, of which the Egyptian desert is a part. Egypt will serve well as an example, especially as it has long been used as a winter sanitarium, and a certain amount of experience of its effects has been accumulated.

The chief features of the climate of Egypt most typically exemplified in that of the desert. are, as shown by the works of Marcet, Sandwith, and Zagiell :

(1) Warmth. The mean temperature of Cairo in winter is $58.3°$ F., the summer mean is $76.1°$ F. ; the maximum at Cairo is $111°$ F., and the minimum as low as $35°$.

(2) Great difference between night and day temperatures, due to radiation, amounting in winter to $23°$ F., or even $38°$ F.

(3) Dryness of atmosphere. The rainfall is small :

at Cairo, 1·22 inch, falling on from twelve to fifteen
days, and is less in Upper Egypt. At Thebes, it is
rare for any rain to fall, and in the province of Esneh
it is almost unknown. The difference between the
wet and dry bulbs sometimes amounts to 24° F., and
the annual relative humidity percentage varies from
58·46 at Cairo to 45 at Luxor. The atmospheric
dryness is proved by the mummies, which remain
unchanged for centuries.

(4) Great atmospheric purity. According to Prince
Zagiell's observations, while ordinary atmospheric air
contains four parts of CO_2 in 10,000 parts, the air of
the desert contains none at all, and putrefaction
appears checked ; meat exposed to the air becomes,
without any trace of decomposition, mummified in
three weeks. Surgeons say that wounds and ulcera-
tions heal rapidly. The climate suffers from very hot
winds, such as the *khamsin* or south-east, which brings
the sand of the desert, and is sufficiently hot to shrivel
up roses and other flowers and to warp and crack
unseasoned wood. Under its influence both natives
and Europeans droop and become listless.

The climate, though warm, is a great contrast to
that of Madeira, being dry and largely influenced by
the results of radiation, instead of markedly equable,
and it has a most beneficial influence on phthisis,
provided the amount of lung area attacked is not
excessive, and there be no fever.

I have notes of 26 consumptives who have tried
this climate for one or more winters, and the results
are favourable. The patients were 23 males and 3

females the average age of the males being 28·43 and that of the females 21·33. The average history of the disease before spending the winter in Egypt was 36·65 months, that is, a little more than three years, thus showing that the cases were not all of early phthisis. Family predisposition was present in 17 cases; hæmoptysis in 18 cases. Seventeen patients had lung tuberculosis alone, and 9 tuberculosis with excavation. In 6 both lungs were implicated, and in 20 the disease was unilateral.

Some patients passed one winter in Egypt, others two, and a few three or more winters, the average length of residence per patient being 6·46 months.

The general results were : improved, 17, or 65 per cent. ; stationary, 3, or 11 per cent. ; worse, 6, or 23 per cent. The local results were arrest in 1 case, decrease of disease in 10, a stationary condition in 3, advance of disease in 4, advance and extension in 3, extension alone in 3, or local improvement in 11 ; a stationary condition in 3, deterioration in 10 : a much worse result than the general.

Dr. F. M. Sandwith's statistics of 104 British phthisical visitors who wintered in Egypt, give 72 improved, 18 stationary, 7 worse, and 7 deaths. Dr. Sandwith gives but few particulars of these patients, except that the fatal cases were all hopeless before arrival in Egypt. He also furnishes an account of 298 natives who were admitted into the Kasr El Aini Hospital for phthisis, many in an advanced state of disease, who died a few days after admission. The interesting feature is the different duration of the dis-

ease among the different races. Among the Egyptians
the average duration of the disease before death was
twelve and a half months; among the Soudanese (a
remarkably fine race), seven and a half months ; among
the Arabians it was still less, and among the negroes
generally the duration was very short indeed. This
is a great contrast to the well-ascertained average
duration of the disease in England which, according
to my[1] statistics among the upper classes, reached
nearly eight years.

This dry climate suits asthma remarkably well, and
I have known many asthmatics who have kept entirely
clear of attacks during a winter on the Nile. In cases
of chronic bronchitis the cough diminishes, expectoration is rapidly reduced, and at last ceases altogether ;
while the climate suits emphysema on account of the
dry warm air and level country, and abundant air
without exercise to be obtained on the Nile steamers
or dahabiyehs. The patients I have found who do best
in Egypt are cases of chronic pneumonia and chronic
dry pleurisy, bronchitis, and chronic rheumatism,
the clearing up of chronic pneumonia and pleurisies,
with the cessation of all symptoms, being remarkable.
One of the best results is the promotion of sleep,
which may be due partly to the cool nights and partly
to the absence of marine influence, which is more or
less exciting. Needless to add, that to profit by the
climate patients should live in the desert, either by
ascending the Nile, or residing at Luxor, or at
Helouan, or even at the Mena Hotel at the foot of

---

[1] *Pulmonary Consumption*, 2nd edition, p. 325.

the Pyramids, as Cairo has all the disadvantages of a large city for invalid residence.

Owing to the increased area of cultivation on the banks of the Nile, that river, in a considerable portion of its course through the land of Egypt, does not represent the climate of the desert, which can be found to greater perfection at some distance from its banks.

I subjoin a good example of the effect of the Egyptian climate.

Case I.—A Greek gentleman, aged twenty-two, consulted me on January 21st, 1877. His father and his brother had died from consumption, the latter of a very acute and obscure form of the disease at twenty-two. He complained of cough and expectoration, but had not lost much flesh, but was somewhat languid. There was no rise of temperature or of pulse. On examining his chest I found slight dulness above the right scapula, tubular sounds above the clavicle, with crepitation on cough, and some tubular sounds above the left scapula, and prescribed cod-liver oil and a tonic. Five months later, after returning from Palermo, where he had been in the interval, I found him suffering from diarrhœa, three to four motions a day, and decidedly thinner, with a reddish tongue, and with increased physical signs. The crepitation was now audible on the right slide from the clavicle to the third rib; distinct crepitation was also heard at the left apex. Under dieting and treatment the diarrhœa subsided and the cough diminished, and the patient improved ; but in consequence of a relative's death, he travelled from London to Leghorn at the end of June, and was there laid up with a feverish attack and a return of the diarrhœa. I heard no more of him until June 20th, 1878, when I was urgently summoned to Naples, and found the patient greatly emaciated and very weak, suffering from profuse diarrhœa, stools varying from four to twelve a day, loose and ochrey, his cough troublesome with num- mular sputa, tongue reddish, evening temperature 100° F., pulse 96, respirations 28. He was scarcely taking any food, but what he did take was solid and of an unwholesome character. On the right side I found tubular sound over the first interspace, on the left side dulness to sixth rib, cracked pot sound in the first interspace, coarse crepitation to the fourth rib and above the clavicle, and distant cavernous sounds in the first two spaces. Posteriorly there was dulness over the upper two thirds, with fine crepitation ; breathing was very deficient in the lower portion of the lung.

The Italian medical man whom I met in consulta- tion informed me that the patient had been suffering from diarrhœa for months, which was with difficulty checked but not stopped, and solid food with claret had been permitted. I found the patient with an excellent *sœur de charité*, who complained that she had no system of treatment to pursue ; a first-rate cook, quite competent to carry out an invalid dietary, but with no directions ; and with a devoted brother and faithful servants longing to be of assistance, and

yet the poor man was lying in the sweltering heat of
Naples in June, starving for want of suitable food, and
such comforts as no hospital patient lacks in England.
I succeeded in removing him to Leghorn in an invalid
carriage, and he bore the journey well, and under the
influence of a nutritious liquid dietary and cooler air
he soon improved, the diarrhœa being at length brought
under by injections of linseed tea. The temperature
and pulse fell, and in the autumn he was sufficiently
recovered to bear a journey, and I recommended his
wintering in Egypt, and ascending the Nile in a
dahabiyeh. He returned to Italy next spring greatly
improved, having gained flesh, with cough and
expectoration reduced, and the physical signs improved, as the doctor in Cairo reported to me. He
spent a second winter on the Nile, with favourable
results, but passed the summer of 1880 at Naples,
where he seems to have led an incautious life, and the
diarrhœa returned. He died in October, and over
his grave his brother has erected a magnificent
reproduction of one of the temples at Luxor as a
tribute to the land of Egypt.

In this case the patient was a southerner ; the tuberculous disease was hereditary and developed rapidly,
and the obstinate character of the diarrhœa and the
nature of the stools passed left little doubt as to the
existence of intestinal ulceration. When I visited him
at Naples his disease was making such rapid progress
that it did not appear probable that he would last
many weeks, but with change of climate and careful

dietetic and medicinal treatment he rapidly improved, and under the dry and warm Egyptian climate he lived on for more than two years in ease and apparently in great comfort. Doubtless the fact of his southern origin had some influence on the climate of Egypt proving so congenial to him.

LECTURE II

Dry Warm Climates (continued)—The Riviera—Three Elements of
Warmth—Southern Latitude—Shelter from Cold Winds—
Mediterranean Sea—Meteorology—Results of Climate in 210
Phthisical Patients—General and Local—Author's Method of
Classification—Four Clinical Examples of Arrest and Improve-
ment—Climate in Bronchitis, Asthma, Chronic Pneumonia and
Anæmia—Contra-indications to Climate—Malaga—Mediter-
ranean Islands—Malta—Corsica—Sicily—Corfu—Algiers—To-
pography of Algeria—Causes of Moist Climate—Varieties of
Temperature—Climate of Algiers—Mountain Stations—Biskra
—Cases of Arrest and Improvement—Tangier—Remarkable
Position and Mild Climate—Results in Two Consumptives
—Southern California—Its Wonderful Fertility—Soil—
Climate—Eastern Portion—Death Valley—Western Portion
—Factors of its Climate—Influence of Kuro Siwo Current
—Equable Temperature—Rainfall—Winds—Santa Ana—
Mists—Case of Arrested Phthisis—Australian Climates—The
Littoral—The Highland—Inland Plains—Paramatta—Illa-
wara—Twofold Bay—Mountain Stations—Riverina—Darling
Downs—Climate—Rainfall—Hot Winds—Southerly Bursters—
Cold Climates of Minnesota and Canada—Their Advantages.
Moisture—Sea Voyages—Opinions of Gilchrist, Walshe and
Maclaren — Rochard — Rattray's Striking Results — Loss of
Weight in Hot Climates—Influence on Skin and Kidneys—
Causes of Unpopularity of Sea Voyages—Clippers to Australia
—Steamers through Red Sea—Dangers to Consumptives—

Temperature and Humidity of Voyage to and from Australia—
Sedative and Tonic Effects—Voyage to Cape—Voyage to West
Indies and South America—Results of Sea Voyages in sixty-five
Phthisical Patients—General and Local—Excellent Influence in
Chronic Pleurisy, Empyema, Chronic Bronchitis and Scrofula
—Hæmorrhagic Phthisis, and Limited Tubercular Cavities—
Neuroses—Arrest of Disease in Three Cases.

THE next warm climate of which I propose to discuss the results, is that of the Riviera, which has been
fully described by Drs. Henry Bennet, Edwin Lee,
Marcet, Sparkes, Burney Yeo, and myself,[1] and its
health and pleasure resorts, being familiar to most of
us, will not require a long notice.

The region extends along the north Mediterranean
coast from Hyères to Spezzia, and owes its warm winter
and spring climate to (1) its southern latitude, lying
between 43° and 44·5° N. lat. ; (2) its protection from
cold winds by the mountain ranges of the Maritime
Alps, and their spurs the Basses Alpes, the Maures,
and Esterels ; (3) the warming and equalising influence of the Mediterranean Sea, which from its tide-
lessness, its salinity, and its high winter temperature
(for it has a warmth of 5·8° to 8·6° F. higher than
the air in winter) exercises a most powerful effect on
the Riviera climate, and nearness to, or distance from
this sea determines the amount of nocturnal radiation.
The winter mean temperature varies from 50·8° to
51·5° F. The minimum, from 42° to 46°, being lowest
in December. Occasionally it sinks to the freezing
point, and snow sometimes falls. A feature of the

Climate of the South of France, 2nd edition.

climate is the rapid fall after sunset, especially at any
distance from the sea.

The relative humidity percentage varies at the
different stations from 61 to 74 per cent., and the
rainfall is 31 inches, distributed chiefly over September, October, and November, the rain coming
down in heavy showers, as much as 4½ inches being
known to fall in nine hours and a half. The rainfall
increases eastwards on the Riviera, and diminishes
westwards.

The principal winds are the north-west, or mistral,
a dry wind, prevailing chiefly in March, and bringing
fine weather, the north-east, or bise, a cold wind, and
the south-east, or sirocco, a warm enervating one.
The westerly winds are dry, and the easterly moist,
the opposite of what obtains in Great Britain and
Ireland.

The winter climate of the Riviera is clear and
bright, with a good deal of wind, but devoid of fog or
mist ; with a mean of 8° to 10° higher than that of
England, with half the number of rainy days, and four
or five times the number of bright ones.

RESULTS OF THE RIVIERA CLIMATE IN PHTHISIS

Before giving the results of 210 phthisical patients
who passed one or more winters in this district. I will
say a few words on my statistics generally. They are
composed partly of cases hitherto unpublished, and
partly of those already communicated to the *Medico-*

Chirurgical Transactions, the latter being brought, as far as possible, up to date.

The duration of life is not calculated, because the object of the figures is solely to determine the exact influence of the climate on the patients during their sojourn at the health resort. Many of these patients went to various climates in succeeding winters. The results are divided into *general* and *local*. The general referring to the functions of digestion and assimilation, circulation, and respiration, to gain or loss of strength and weight, and to the main symptoms, including cough and expectoration. The categories of *improved*, *stationary*, and *worse* speak for themselves, and by *cure* is meant the disappearance of all symptoms of disease, and the restoration of the patient to apparent health. The local results are classified under the headings of " Arrest of Disease," where, in the case of first stage, or tuberculosis patients, all physical signs have disappeared, and in softening and excava-tion cases it signifies signs of contraction and entire disappearance of cavernous sounds. " Decrease of disease" means simply a diminution of the physical signs either in area or intensity. " Advance of Disease" means the process of softening and excava-tion. " Extension " and " Advance and Extension " and " Stationary" all speak for themselves.

The examination of sputum for tubercle bacilli has been carried out in most of the cases since 1882, the year of Koch's discovery of the bacillus. The ex-aminations have been made generally for the sake of diagnosis, no great importance being attached to its bearing on prognosis. A large number of the patients were treated before the era of bacillus discovery. Examination for lung tissue has frequently been carried out. The number of deaths it has been found impossible to ascertain, because many of the old patients have been lost sight of. In the more recent cases it has been ascertained, but to furnish this alone, without that of earlier cases, would only be misleading.

One hundred and forty-nine patients were males and sixty-one females, the average age of the males being 29·53, and of the females 26·31. Family predisposition was present in 117, or 55·6 per cent., and hæmoptysis in 124, or 59 per cent.

The average history of tuberculous disease before wintering on the Riviera was—for males 26·46 months, and for females 26·11 months. The cases were those of chronic phthisis, some intermingled with inflammatory conditions, some of the hæmorrhagic type, and more of the scrofulous form.

The condition of the lungs of the patients has been given in summary in Table I., which shows that 123, or 59 per cent., were cases of tuberculosis without excavation, and that in 87, or 41 per cent., softening and excavation had taken place. Of the first stage cases, 55 had the right lung affected, 28 the left lung, and in 40 both lungs were attacked. Of the excavation cases there were right lung cavities in 23, right lung cavities with left lung infiltration in 8, left lung cavities in 29, and left lung cavities with right lung infiltration in 20. Seven patients had both lungs undergoing softening or

E

excavation. Summing up, we may say that the disease was unilateral in 135, or 64 per cent., and bilateral in 75, or 36 per cent.; and that 59 per cent. had different amounts of consolidation, and 41 per cent. had cavities.

The 210 patients were scattered about the Riviera stations, 76 went to Hyères, 43 to Cannes, 25 to Nice and Cimiez, 36 to Mentone, and 23 to San Remo. Twelve visited various stations on that coast, 27 wintered in the south of Europe, being part of each winter on the Riviera, and many passed one winter at one health resort and the next at another, while some remained faithful to one station and passed five or even ten winters in the same place. According to our returns, more time was spent at Hyères than any other place. The average length of residence was nine months.

The *general* results of these cases showed 6 cures, 131 improved, making, with the cures, 65·23 per cent. of improved, 21, or 10 per cent., stationary, and 52, or 25 per cent., worse.

The *local* results (Table I.), excluding unknowns, were as follows : 12 instances of arrest (nearly 6 per cent.) ; 62 of decrease, making 74 improved, or 37 per cent.; 36 stationary, or 18 per cent.; 35 cases of advance, 27 of advance and extension, and 30 of extension, making in all 92 to be classed as worse, or 45 per cent.

This total cannot be considered very satisfactory for either category. The percentage of 45 of worse is high, but we must bear in mind the large proportion of excavation cases and of bilateral affections. If we examine the results of each class we see the first stagers improved more than the cavity cases in the proportion

TABLE 1.—*Showing the Condition of Lungs of 210 Cases of Consumption who Wintered on the Riviera and in the South of Europe.*

Stage.	Number.	Percentage.	State of Lungs of those Wintering on the Riviera, etc.	Arrest of Disease.	Decrease of Disease.	Stationary.	Advance of Disease.	Advance and Extension.	Extension.	Unknown.	Improved.	Stationary.	Worse.
Tuberculisation, 1st stage	123	59	55 had right lung alone affected	5	22	9	9	6	1	3			
			28 had left lung alone affected	1	10	2	4	5	6	—			
			40 had both lungs affected	2	9	8	11	5	5	—			
				8	41	19	24	16	12	3	41	16	43
Softening and excavation (2nd and 3rd stages) + tuberculisation.	87	41	23 had right lung in 2nd or 3rd stage	3	8	5	—	1	5	1			
			8 had right lung in 2nd or 3rd stage, and left in 1st stage	—	1	2	1	3	—	1			
			29 had left lung in 2nd or 3rd stage	—	8	5	4	3	7	2			
			20 had left lung in 2nd or 3rd stage, and right in 1st stage	1	3	5	3	2	5	1			
			7 had both lungs in 2nd or 3rd stage	—	1	—	3	2	1	—			
				4	21	17	11	11	18	5	30	21	49
				12	62	36	35	27	30	8	37	18	45

	Per Cent.			Per Cent.
Both lungs affected in	75 = 35·7		Right lung alone affected	78 = 57·0
One lung affected in	135 = 64·3		Left lung alone affected	57 = 43·0

of 41 to 30 per cent., and deteriorated less in proportion of 43 to 49. We also note that in the first

E 2

stages the unilateral improved more than the bilateral
as 45 to 27, or 5 to 3, and the right-sided cases pro-
gressed better than the left-sided ; and the same held
good with regard to the cavity cases, the right cavities
showing greater progress than the left. When we in-
vestigate the individual cases of the above statistics,
where the disease improved and even underwent arrest,
we cannot explain why some patients prospered so
much more than others.

The following cases are examples of the beneficial
effect of this climate :

Case II.—In October, 1875, a clergyman, aged
thirty-nine, consulted me with a three months' history
of phthisis following overwork. He had slight pyrexia,
had wasted considerably, and I detected a cavity in
his right upper lobe. He passed one winter at
Bournemouth, taking tonics and cod-liver oil, but
with no local or general improvement ; and spent
the following winter on the Riviera with great
advantage, returning without cough or expectoration,
and having regained his strength and weight. The
signs of cavity were no longer present, and no physical
signs beyond dulness and flattening remained. That
was in 1876, and since that date he has been a hard-
working clergyman in the midland counties, out in
all weathers ; and though occasionally under my care
for the results of over work, he has had no recurrence
of lung disease, though eighteen years have elapsed
since his return from the Riviera.

Case III.—A young man aged twenty, with a
decidedly phthisical family history, who broke down

after overworking for a competitive appointment, and after four months of tuberculous symptoms, presented a cavity, with extensive infiltration, of the right lung. He spent the winter of 1873-4 at Hyères, lost his cough and expectoration, gained 2 stone in weight, and when he returned to England I could detect no signs of a cavity, but only dulness and slight flattening of the side. This gentleman held an appointment in the Army Transport Corps during the Zulu war, and distinguished himself, but afterwards contracted typhoid fever in the Transvaal, and died at the end of 1879.

These patients may be considered fair instances of the arrest of tuberculosis in the excavation stage, and there are many of the same type to be found among the Riviera patients, some of whom are members and ornaments of the medical profession, who have not only recovered, but have struck root on the Riviera, and practised their profession for many years with advantage to their fellow countrymen and credit to themselves. Such cases as those of the late Dr. Henry Bennet of Mentone, and Dr. Griffith of Hyères, are excellent examples of what energy in a fine climate may do in arresting tuberculous disease.

Where this climate does not arrest disease, it may prolong life for many years, as in the subjoined case.

Case IV.—A young lady, born in Brazil, though of Scotch parents, was seen by me September 6th, 1876, with a history of consumption of six months' standing, and tuberculosis of both upper lobes and of the left lower lobe. She had wasted considerably, and only weighed six stone. Her breath was short on exertion, her cough very troublesome, but there was no pyrexia. I had considerable hesitation in recommending her to winter on the Riviera, doubting her fitness to travel; but fearing the English winter for so frail a patient brought up in the South, I determined to make the venture, which turned out a successful one, and advised Hyères. The case was unfavourable from the first, but she returned from Hyères greatly improved, with a cavity in one lung and a gain of 1 stone 6 lbs., and the cough and expectoration much reduced. She returned to Hyères, and passed altogether nine winters there. For three or four years the disease remained quiescent, but about 1880 there was a distinct increase in lung disease, and intestinal ulceration came on. She rallied, however, and continued to spend her winters at Hyères and her summers at Leamington, till she died at the latter place in the summer of 1885, nine years after I saw her first. This patient undoubtedly owed much to the climate of Hyères, especially to the sunshine, in which she revelled, as was not unnatural for a daughter of the South. The changes in her weight were remarkable. She was of medium height, but of a slender figure, and her weight would at times fall to 5 stone 10 lbs., and then increase up to 7 stone again, this undoubtedly depending on the amount of food consumed and the regularity of taking cod-liver oil.

The winter climate of the Riviera is not only clear and bright, but it is intensely stimulating, especially to the nervous system, and consequently is contra-indicated in all neuroses, where the sedative element

is required. Want of sleep is a very common
complaint on the coast, and the general remedy is
to go a few miles inland—to Cimiez or Grasse, for
instance—where the Mediterranean influence is not
so strong.

The cases which appear to do best on the Riviera
are : (1) chronic bronchitis and emphysema : (2) chronic
pneumonia with or without bronchiectasis ; (3) bronchial
asthma ; (4) phthisis, in which inflammatory attacks
have been the predisposing causes of the disease ;
(5) scrofulous phthisis ; (6) unilateral first stage, which
improves far more than bilateral phthisis. Tuberculisation cases generally improve more than cavity
cases ; (7) anæmia.

I have seen many cases of chronic pneumonia of
some standing clear up with marvellous rapidity.
Asthma does fairly well on the Riviera, and best of
all at Hyères, partly on account of this climate being
less stimulating than that of other resorts, and partly
because the valley is a more open one than some of
the others.   For all cases, either phthisical or non-
phthisical, where any degree of fever is present the
Riviera is contra-indicated, and experience points to
its stimulating character increasing, instead of decreas-
ing, the fever.   It is also contra-indicated in insomnia
and in most nervous affections, specially hysteria, on
account of the exciting qualities of the climate.

It may be well here to give a passing notice of
other Mediterranean stations possessing climates
similar to that of the Riviera, though they present
some contrasts.

Malaga on the south coast of Spain, occupying a
well-sheltered position, enjoys a winter climate warmer
and drier than the Riviera and apparently with smaller
daily and seasonal range, and were it situated in
France or Italy, would probably become a popular
health resort on account of its fine climate, but
Spanish authorities and Spanish innkeepers do not
manifest the same readiness to meet the requirements
of invalids shown by the French and Italians, and
hence the neglect of a most promising sanitarium.
All the patients (ten) whom I have sent to winter
there, have done well, in spite of the various drawbacks.

Certain islands of the Mediterranean, such as
Malta, Corsica, Sicily, and the Ionian Isles have long
been utilised as sanitaria for lung disease. The
climates of these must vary greatly according to
latitude and shelter from winds, but with the exception of Malta, they may all be said to enjoy a warmer
and a moister climate than the mainland. This for
British patients generally is undesirable, as most of
these are exiled from their native land on account of
its dampness; but there are some who sleep sounder
at Ajaccio and at Palermo than on the Riviera, and
are beneficially affected by a transference to one of
these localities. In Sicily, the sheltered east coast,
with Catania and Aci Reale on the slopes of Etna, and
the lovely Taormina overlooking the Straits of
Messina, may be mentioned as fine winter climates of
the insular type, and well adapted for patients who are
able to take walking exercise. Ajaccio, in Corsica, is

an excellent winter and spring shelter, rather milder and moister in climate than Mentone, with a larger rainfall. Here we find beautiful walks and drives. In the summer and early autumn there is an occasional risk of malaria, but then all invalids are away. It has always been a matter of surprise to me that Ajaccio has not been more utilised as an alternative climate by the Riviera medical men, when their own air has proved too stimulating or too marked by radiation extremes, for this mild, moist atmosphere, with its freedom from all but sea breezes, and its good hotels and quiet surroundings, seems to supply the requisite and beneficial change.

Corfù, in the Ionian Islands group, enjoys a beautiful climate and is a perfect garden of luxuriant vegetation, with the advantage of fine roads and fair hotel accommodation, but is subject to storms, and hardly possesses the bracing element sufficiently for our purposes.

Far more important, and more largely frequented by invalids than the Mediterranean islands, is the French province of Algeria, situated between 32° and 37° north latitude. The province extends 3,200 miles from east to west, and 200 miles from north to south, and reaches into the Great Sahara, being intersected by mountain chains, thus presenting a series of remarkable climates which the advance of French civilisation, bringing railways and good hotels, has rendered available and accessible. The Atlas Mountains which rise to a height of 7,000 ft., traverse Algeria in three chains running east and west, the

Lesser Atlas, the nearest to the Mediterranean
Sea, being separated from the Middle Atlas by the
fertile valley of the Cheliff, and the Middle Atlas
from the Great Atlas by the Algerian desert, an
elevated plain containing salt-water lakes.   South of
this last chain stretches the Great Sahara desert, which,
according to the late Dr. Henry Bennet, is the key to
the Algerian climate, and converts what would be a
dry climate into a moist one. The   atmosphere over-
lying this rainless tract of desert becoming heated both
in winter and summer, must rise into the higher strata,
and thus a vacuum is formed, which the cooler air from
the Mediterranean basin rushes in to fill, being sucked
in over the summits of the Atlas ranges.   Consequently
the regular winds are, and must be, north-east and
north-west, and south winds only blow exceptionally,
though the sirocco, when it blows, is a terrible blast
from the desert.   These northerly winds, coming from
the Mediterranean or Atlantic, are laden with moisture,
and striking the Atlas ranges are at once cooled and condensed and deposit their moisture in the form of frequent
and abundant rain over the entire Algerian region,
reaching into the desert 250 miles from the sea.   Consequently the rainfall at Algiers is heavy : 32 inches
distributed over 87 days, occurring principally in winter,
the largest rainfall and the greatest number of rainy
days occurring in November, December, and January,
which is a decided drawback for invalids.   During dry
seasons plagues of locusts invade Algeria from the
desert, entering by the passes and causing wholesale
destruction of cereals and vegetation generally.   The

rainfall increases on proceeding eastwards, the province
of Constantine, the most easterly, having the largest,
probably because of its being most wooded, and the
westerly province of Oran, where there are few or no
forests, has the smallest rainfall. The mean winter temperature is 56° F., and the difference between winter and
spring is small. The climate is milder and moister
than that of the Riviera, and occupies an intermediate place between it and that of Tangier
(Morocco), where the equalising influence of the
Atlantic is more felt in moderating the extremes : but
the great recommendation of Algiers lies in its
numerous and diverse sanitaria. Algiers, with its
suburbs Mustapha Supérieur and Inférieur, offers
shelter and saline breezes ; Blidah, Medeah and
Milianeh are excellent mountain stations, and the air
of the desert can be tried at Biskra, in the desert, now
connected with Algiers by railway. Hammam R'Irha
combines moderate elevation in a well-sheltered valley
with hot springs. Few districts can show such
admirable spring resorts, and the only drawbacks seem
to be the amount and period of the rainfall and the
distance from this country.

Of the dozen consumptive patients of whom I
have notes, who have wintered once or oftener at
Algiers, the large majority improved greatly, and
number at least two cases of arrest, but I note that
the greatest improvement took place where patients
resided in villas with gardens, and not in hotels. In
one case, where a young lady, a member of a very
consumptive family, developed the disease and a con-

siderable cavity had formed in one lung, complete
contraction of the cavity took place with arrest of the
disease in two winters, and the lady has since married
and has resided for the last nineteen years in England
without any signs or symptoms of relapse.   Another
lady with well marked tuberculosis of one lung spent
two winters in a villa at Mustapha Supérieur with the
result that the disease became arrested, and since that
date she has been able to pass twelve winters in
Scotland with impunity.

Tangier has long been noted for its remarkable
climate, which apparently combines the warmth of the
Mediterranean, with the equability of the Atlantic, and
being separated by a series of mountain ranges from
the Sahara desert, does not share all the features of the
Algerian climate, though its rainfall is large—thirty
inches—occurring chiefly in October and November.
The mean winter temperature is about 60° F , and the
diurnal variations are slight.   The climate is mild, with
a bracing element, owing to Atlantic winds, and its
influence on chronic phthisis in the few cases I have
watched, has been eminently beneficial. In two cases the
results were most striking. In an Indian chaplain, aged
thirty, phthisis had supervened on several attacks of
intermittent fever and a cavity quickly formed, the
opposite lung also becoming involved and the condition was gradually deteriorating in this country, when
I sent him first to Hyères and then to Tangier, and at
the latter place he resided five years, gradually improving in health and for at least three years held the
office of chaplain.   He died from an attack of diarrhœa.

This was a case in which acute symptoms seemed
likely to terminate in a few months at home, but the
fine climate apparently transformed the character of
the disease from acute into chronic.   Another case was
that of a gentleman, aged thirty, who had lost a brother
and sister from consumption and had another brother
similarly affected.   He was attacked himself at the
age of twenty-seven, and after spending two years in
Cyprus, where he deteriorated, he was appointed
Consul at Tangier and rapidly improved.   The
disease gradually became arrested, and though he had
a severe attack of scarlatina no return of his symptoms
took place.   When last heard of, in 1894, he was still
living at Tangier, in fine health, with active habits,
though thirty-six years had elapsed since he was first
attacked with phthisis.[1]

SOUTHERN CALIFORNIA

Another dry, warm climate, which must not be
overlooked, is that of Southern California, which
somewhat resembles that of the Riviera, but is
warmer and more equable. Its meteorology has
already been alluded to in the first lecture, and in
my tour through the United States in the autumn of
1892 I visited this region and carefully examined
many of its towns and health resorts

The Tehachapi mountains, running east and west

[1] Curiously this gentleman had two sons born at Tangiers, who
both became consumptive.   One left Tangiers and died, and in the
other, who has always resided there, the disease has become arrested.

across California, join the coast range and the Sierra Nevada, and what is called Southern California is the part of the State lying south of the Tehachapi, between the Sierra Nevada mountains and the Pacific. It is an irregular rhombus of country, situated between latitude 32°30′ and 35°40′, with a coast line of 330 miles. The western portion is traversed by the coast range of mountains, and encloses numerous beautiful and fertile valleys, and then widens out into a large plain, limited to the east by the Sierra Madre and San Bernardino ranges. The soil of this plain is composed partly of adobe or red clay—of which the picturesque old Spanish convents and mission-houses are built—and partly of granite detritus. It lies at some height above the sea level, is protected from all cold blasts, and watered partly by streams and partly by irrigation canals, which the Californians have constructed with great skill and energy, and it is now one of the most fertile spots in the world. The fruits of both tropical and temperate regions are produced here, and while you see gigantic oranges and lemons, grapes, figs, and olives of the south of Europe, persimmons of Japan, guavas, loquats, custard apples, and pineapples of the tropics, you also recognise apples, pears, peaches, apricots, nectarines, mulberries, quinces, blackberries, raspberries, currants, and nuts of our own climate, so that a South Californian dessert is very cosmopolitan. Strawberries are to be had all the year round. Vegetables are in the same variety and profusion.

This plain, with the different valleys opening on

it, and the western slopes of the Sierra Madre and
San Bernardino mountains, constitutes a vast garden
of hundreds of miles in extent, and supplies vegetables
and fruit to the United States all the winter through,
and also exports tinned fruits largely to all ports of
the world. Here is situate Riverside, with its famous
magnolia avenue and orange groves.

The eastern portion of Southern California, north
and east of the San Gabriel and San Bernardino ranges,
consisting of the Mojave and Colorado deserts, lies at
an altitude of from 1,000 to 3,000 feet. The soil, being
chiefly the detritus of granite, is good, but the barren-
ness is caused by want of water. The rainfall is very
small indeed, being diverted by the mountains to the
west, which condense all the moisture coming from
the Pacific, and nearly all the streams of the San
Bernardino range flow to the Pacific, and not eastwards. At present only cactuses and gigantic yuccas,
or what are called yucca palms, grow here ; but parts
of the Mojave desert have yielded good crops under
irrigation, which is carried out by tunnelling the
mountains, and it is likely that the whole will in time
be brought under cultivation.

This tract has some remarkable depressions in it,
which are supposed to have at one time formed part
of the Gulf of California, and one of these, called the
Arroyo del Muerte, or Death Valley, is, in fact, below
the sea level. The name was given after a party
of emigrants perished there from thirst in 1850. It
lies at the northern margin of the Mojave desert, about
200 miles from the coast, between the Funeral and

Amargosa and Parramint Mountains, and is about 75 miles in length and 12 to 25 in width. This valley was formerly an old bitter-water lake, and is now covered with a crust of salt and borax deposits, and devoid of all vegetation. The interesting point about the valley is its meteorology, it being one of the hottest and driest places in the world. Professor Mark Harrington, the Chief of the Weather Bureau, in an able report on this climate in summer, states that the temperature occasionally rises to 122° F., and has been known to reach 137°. It rarely falls in the five hot months below 70°, the average summer mean temperature being 94° F. It is not only hot, but, being surrounded by high and bare mountains, persistently hot, during the five summer months, and the heat is increased by the warm blasts from the desert to the south. The air is in active motion, and there are often sandstorms. The rainfall is 0·13 inch, the relative humidity is 29, the mean daily range 29°. In winter it is cold at times, and the radiation excessive. As a rule, all travelling through the valley must be done at night in summer ; and there is a story of men becoming insane who have been exposed to noonday heat, and that one driver taking a load of borax, in which a brisk trade is carried on, died with the water canteen in his hand. The cause of this great heat and aridity is the position of the valley, the sandy soil and wind adding to the dryness. There is nearly as much heat at Fort Yuma, further south.

However, it is not this portion of South California which is fitted for health stations, but the western part fringing the coast; and here, in the plain of San Gabriel and on the slopes of the beautifully wooded ranges, we find plenty of charming sites for invalid residence. Los Angeles, with pleasant villas and gardens; Pasadena, with its orange and lemon groves and its fine hotel; Sierra Madre, with its comfortable hotel *pension*, its vineyards, and extensive views from the hillside; and, on the coast, Santa Monica, where there is an ostrich farm; Santa Barbara, a well-protected site; and San Diego, with the beautiful Coronado beach, close to the Mexican frontier. At the last place I saw greater profusion of tropical vegetation than ever before. Also in North California, south of San Francisco, is Monterey, on the Pacific, a well-equipped station, with an excellent hotel, and good drives in the splendid pine woods. All these places have hotels, often good ones, and if invalids insisted on certain requirements they might be much improved; but at the present moment the Southern Californians do not understand that in their cities the unceasing noise of cable and electric cars detract from the conditions of repose needful for recovery; and more elasticity of rules and more attention to comfort are desirable even in the palatial hotels of which they are proudest.

The climate of Southern California depends for its warmth and equability on three factors:

(1) Its southern latitude.

(2) Its protection by the various mountain ranges from cold northerly or easterly winds.

(3) The influence of the Pacific, and especially of

the warm Kuro Siwo, or Black Japan Current, which washes the shores of the Western states.

All these are potent influences, but to the last is mostly due the warmth in winter and coolness in summer.

The Pacific Ocean off this shore, according to Dr. Orme, of Los Angeles, to whom I am indebted for much information, varies in winter temperature from 60° to 70°, and the degree of equability enjoyed by each of the health resorts depends on nearness to, or distance from, the sea.

The annual mean temperature (11 years) of Los Angeles is 61° F. For the winter months (December, January, and February), 54° F. For the summer (June, July, and August), 68°—the maximum 105°, the minimum 28°. On the coast these figures give less extremes.

The rainfall occurs in the winter months, and ranges from 10·82 inches to 16·92, and the number of rainy days from 20 to 30 in the year. The relative humidity is from 60 to 75 on the coast, and diminishes on nearing the inland desert.

The winds are during the summer from the north and north-west; in the winter from the south and south-west, the Pacific Ocean being the principal source. There is one westerly wind called the chinook, or Santa Ana, which is hot and disagreeable, and during its prevalence the air becomes highly electrified. Horses' tails stand out, and the hair crackles when stroked. The great drawback of South California are the mists which roll in from the

Pacific, sometimes quite suddenly in the middle or after-part of the day, and give a very raw feeling to the air. They are worst on the coast line, but extend many miles up the valleys, though they are said never to reach an altitude of 1,000 feet.

The strong point of the climate is equability, which gives it an advantage over most other warm regions like Egypt and the Riviera ; for in South California a patient can live with comfort all the year round, the difference between the seasons not being accentuated. It is moister than its neighbours, Arizona, Utah, and Colorado, but still by no means a damp climate, and it allows of an open-air life, and of camping-out to any extent. The country is well suited for an invalid with a little capital to settle in and find profitable and healthful occupation in fruit or vegetable growing, for which there seems an unlimited demand, and, though the climate is by no means bracing, its equability, combined with dryness, prevents catarrhs and their consequences.

Case V.—I met a medical man in consultation in Bromley, in Kent, in July, 1884, and, on my remarking that he was apparently in a worse condition than his patient, he informed me that he had had hæmoptysis (ℨij to ℨiij) sixteen days previously, and cough and expectoration for some months, with wasting. He had lost his mother, brothers, and a sister from consumption, and had been seriously thinking of leaving England for his health. He asked me to examine him, and, on doing so, I found at the right upper chest crepitation in the first interspace, with some dulness to the third

F 2

rib. I advised his emigrating to South California. A few months later I received a letter from him in South California announcing his arrival, and I heard nothing more till in November. 1892, Dr. Orme brought him to my hotel in Los Angeles, looking very well. He then told me that he had brought his wife and family out, and had gradually improved in health, and, losing his cough, he had settled in practice some ten miles away, on the slopes of the Sierra Madre, and had been able to lead an active life and support his family ever since. He had some slight hæmoptysis last year, but, with that exception, had been well. The only physical signs were some tubular sounds in the first interspace on the right side. He spoke in high praise of the climate, specially of the lower ranges of mountains, which are always cool and pleasant and free from fogs.[1]

According to Dr. Davidson,[2] of Los Angeles, the cases which particularly do well are : (1) Chronic phthisis in all stages if free from fever, and with sufficient lung space to allow of exercise. (2) Phthisis with bronchiectasis. (3) Catarrhal phthisis, and phthisis with a tendency to inflammatory attacks. Chronic bronchitis and asthma do well here.

Another dry, warm climate, which has been long used for the treatment of lung disease is the Australian, which, owing to the size of the continent, extending from 10° to 40° South latitude, comprises great varieties

---

[1] I regret to have to report that this gentleman died in July, 1893, after an attack of influenza.

[2] "On the Climate of South California and its Suitability as a Health Resort," *Glasgow Medical Journal*, June, 1893.

of climate, from temperate to tropical. Three climatic regions of Australia have been enumerated by Dr. Lindsay[1]: (1) The Littoral; (2) the Highland; and (3) the region of the Inland Plains. The Littoral, where lie the principal cities, consists of a narrow strip of country, varying from 30 to 150 miles in breadth, between the ocean and the mountains, and suffers from being open to two evils, cold blasts from the Antarctic Circle and hot winds, (the characteristic of Australia), blowing from the central inland desert. This region cannot be generally recommended to invalids, though certain localities, such as sheltered spots like Paramatta, Illawarra, Eden and Twofold Bay are exceptions to this rule. The highland region, embracing the slopes of the Australian Alps, which rise to 7,000 feet in Mount Kosciusko and the Blue Mountains, stretches from Queensland to South Australia; but although it presents a choice of elevation from 2,000 to 7,000 feet, yet with the exception of Mount Macedon in Victoria, and Mount Victoria in New South Wales, it offers but little accommodation for invalids. On Mount Macedon, a fine sanitarium, Braemar, Woodend, has been built on a plateau 2,500 feet above sea level, and connected by rail with Melbourne; some accommodation is also to be had on Mount Victoria, but otherwise Australian mountain sanitaria are still in the future. The region of inland plains comprises the vast Riverina plain, a great pastoral district, which has for its boundary on the west the central desert, on the south the Murray

---

[1] *Climatic Treatment of Consumption.*

river, on the north Queensland, and on the east the
Darling Downs, and is the centre of the sheep farm-
ing industry.   The climate of the Riverina is hot,
rising occasionally to 110° F. in summer, though the
heat is not much felt, except during the hot winds.
It is also dry, the rainfall varying from 5 to 24
inches. with an average of 14.   Like the South
Californian climate, the summer heat seems easily
borne, on account of the dryness, and the winter
only consists of a little morning frost, while autumn
and spring enjoy (Lindsay[1]) the ideal perfection of a
climate.   There is no want of accommodation for
invalids in the well-appointed towns, such as Deniliquin,
and also in the various sheep farms scattered over the
region, in many of which a young man, affected with
early phthisis has worked out his cure by adopting a
pastoral life in this healthy district, which is now
rendered available by rail connections with Sydney
and Melbourne.   The Darling Downs in Queensland
enjoy the advantage over the Riverina of an altitude
of 2,000 feet. and of being a considerable distance
both from the desert and the sea, and partaking
largely of the genuine inland Australian climate.   The
towns of Toowoomba and Warwick supply the needful accommodation for patients.   The rainfall in
Australia generally decreases from the coast towards
the central desert, and the fine climates are to be found
not on the coast line, but in the interior.   Another
feature are the winds, of which the hot wind, the offspring of the desert, is a terrible scourge, withering

[1] *Op. cit.*

vegetation, but not greatly affecting human life, and occurring about seven or eight times in a summer. The southerly bursters produced by the in-draught towards the heated interior are of great velocity attaining ninety miles an hour and upwards. These are the bearers of rain, generally ending in storms. Invalids appear to suffer more during the Australian summer than the winter, but stand both well, if living in the Riverina and Darling Downs, and carefully avoiding the cities, which on every ground are unsuitable residences for patients.

Unquestionably the inland climate of Australia is highly beneficial for early phthisis, and can be strongly recommended to more or less vigorous patients with pastoral tastes, who are prepared to spend years in the recovery of their health.

The effect of *cold* climates in the treatment of phthisis and certain lung affections was at one time extensively tried in the State of Minnesota, and many Americans wintered at St. Paul's, Brainherd, and other places with some benefit, though cited results of this climate are not very striking. It is generally advisable to combine some other element, such as altitude, with dry cold.

Nevertheless, a climate like that of Canada, in which outdoor life is combined with extreme cold, and necessitates a large amount of exercise, has generally proved most beneficial in chronic consumption, and it is possible that cold in addition to bracing the muscular and nervous systems, may retard and prevent the multiplication of pathogenic organisms, and thus exercise a powerful antiseptic influence. I have noted that in the Alpine mountain-stations patients generally improve more in winter, when the ground is covered with snow, than in summer, when the soil teems with plant and insect life. Great cold is probably aseptic, but its bad effect lies in reducing the production of red corpuscles and hæmaglobin in the circulation, as was shown by Drs. Bristowe and Copeman's interesting experiments.

MOISTURE

Hygrometry, or the measurement of moisture in the atmosphere, is of as great importance as the measurement of heat, and the distribution of rainfall and the amount of relative humidity are of the deepest interest to medical men and their patients. We all know the refreshing effect on our sensations of rain falling after a long period of dry heat, when we seem to breathe more freely. On the other hand, it would appear that excessive atmospheric humidity prevents free evaporation, though it often promotes expectoration, and is therefore useful in bronchial affections. I have not many instances of the effects of an increase of humidity on the circulation, but the following is a striking one, which occurred to a patient of mine.

A gentleman, aged thirty-two, who had a well-marked cavity in his right lung, and some tuberculisation of his left, was trekking in the Kala Hari Desert of South Africa. The climate was exceedingly dry, and a difference of 25° F. had been noted between the wet

and dry bulbs. A thunderstorm came on, and heavy
rain fell, and the patient immediately had an attack of
hæmoptysis, having been free from that symptom for
years previously.

We have already considered the effect of dryness
as an element of climate, and have taken Egypt and
the Riviera as examples of dry climates.

SEA VOYAGES

The best example of a moist climate is that of the
ocean, as it combines moisture with a saline atmosphere, and in the case of many sea voyages, with a
well graduated rise or fall of diurnal temperature.
The use of sea voyages in the treatment of lung
diseases has long been advocated by Gilchrist.
Maclaren, and others, and Dr. Walshe[1] gave it as his
opinion "that a sea voyage, especially in the case of
young adult males, will occasionally work more
effectual change in the phthisical organism than any
other single influence, or any combination of influences
with which he was acquainted." On the other hand.
Rochard showed, from the records of the French
Navy, that phthisis was more common in the navy
than in the army, and that, with rare exceptions, the
course of phthisis was more rapid on board ship than
on land. This last may be accounted for by the fact
that at the time of Rochard's statistics the French
navy cruised almost exclusively in the torrid zone.
Rattray's[2] important observations on the crews of the

[1] *Diseases of the Lung*, 4th edition, p. 655.
[2] *Proceedings of the Royal Society*, 1869—72.

Royal Navy during voyages in different climates tend
in the same direction, and show that the tropics should
be avoided by natives of colder and temperate
climates, especially by the young, the weakly, and
those suffering from chronic disease. Also that
healthy adults should not remain too long in the
tropics, and should quit them at once if strength and
flesh begin to fail. The most striking result of
Rattray's observations was the loss of weight under
the combination of the tropics and salt diet in able-
bodied seamen. This occurred in 81·55 per cent.,
the average loss being 4 lbs. When muscular exertion was added, the percentage of loss exceeded
91. These sailors in 104 days lost on an average
7 lbs., the shipboys not losing so much as the men.
A wonderful improvement took place on reaching
temperate climes. Rattray found that in the tropics
the urine of the crews decreased $59\frac{1}{2}$ to 42 per cent.,
and the perspiration increased from $8\frac{1}{2}$ to 30 per cent.
There was a diminution of $4\frac{1}{2}$ per cent. of the fluid
exhaled by the lungs, and a slight increase in that
secreted by the bowels. The kidneys appeared to
be the principal eliminators of surplus water, both
in the tropics and in temperate regions ; but the skin
was most worked in the tropics, and the lungs in
temperate climates.

Though the above results are taken from a number of
healthy men and boys, for the most part hard worked
in the tropics, and therefore somewhat differently cir-
cumstanced from our invalids, the evidence is in
favour of avoiding the tropics as far as is possible

in sea voyages. Before the use of high altitudes in
the treatment of consumption, a sea voyage was held
to be the most effective climatic agent; but of late
years the number of invalids who have recourse to
them has greatly diminished, and this for several
reasons. The Australian or New Zealand voyage is
the best climatologically, because it affords the longest
spell of marine influence, and especially if a clipper be
chosen in preference to a steamer. The best route is
round the Cape of Good Hope.

The number of clippers has been greatly reduced of
late years, and accordingly the choice is more limited.
In this age of hurry and scurry many people object to
being shut up in a vessel out of sight of land for three
months at a time, passing a life which must be to a
certain extent monotonous, and occasionally liable to
being put on short commons or salt provisions if the
voyage be prolonged by adverse conditions. About
sea voyages in general it may be said that if the
weather be fine, and the patient able to remain on
deck, the circumstances are very favourable, as under
no other conditions could he procure so much fresh
air; but, in stormy weather, confined to his cabin,
which is always small, often unventilated, with port-
holes closed and hatches battened down, a different
state of things prevails, and even in moderate weather
he cannot be always on deck, and seldom gets
sufficient exercise: for the deck walking, except to
the regular seaman, is apt to become very monotonous.
He may, like too many, spend the greater part of the
day in the smoking room, drinking spirits and betting

on the speed of the vessel, and at the end of the voyage his state may be rather worse than at the beginning. Even sea sickness, though in most cases a passing ailment, does not improve matters. One reason for the decline in the use of clippers is the introduction of steamers, and particularly the increase in the traffic through the Suez Canal and the Red Sea, which of course makes the transit to Melbourne or Sydney a matter of five weeks, very different to the old three months of sailing vessels. Even of the clippers which are left, but few go and return by the Cape of Good Hope, as, for navigation reasons, the return by Cape Horn is preferred; but not so for the patient's sake, as I showed in my Lettsomian Lectures before the Medical Society, for passing Cape Horn necessitates nearing the Antarctic circle, and a very chilly atmosphere prevails for nearly a month. The meteorology of an outward-bound voyage round the Cape from England to Australia in a clipper starting in October is briefly the following, as is shown by a diagram constructed by Mr. John Curtis, of the Meteorological Office from the log of a clipper courteously placed at my disposal by Lieutenant Baillie, the Marine Superintendent.

The temperature ranges from 53° to 58° F. for the first five days. Off the Azores it rises to 60°, and on passing Madeira to 69°. In crossing the line the maximum, 82°, is attained, but breezes are present and temper the heat; afterwards it gradually falls; in 30 south latitude 70° is the average, and this sinks to 58° on reaching the Cape. After rounding the Cape the

temperature, owing to the mixture of the warm
Agulhas current with the Antarctic current, is un-
certain, varying from 47.5 to 56°, the currents over-
lapping each other, and causing great varieties of
atmospheric temperature. The vessel reaches 43 south
latitude and steers eastward, and, on account of the
influence of the Antarctic circle, the temperature

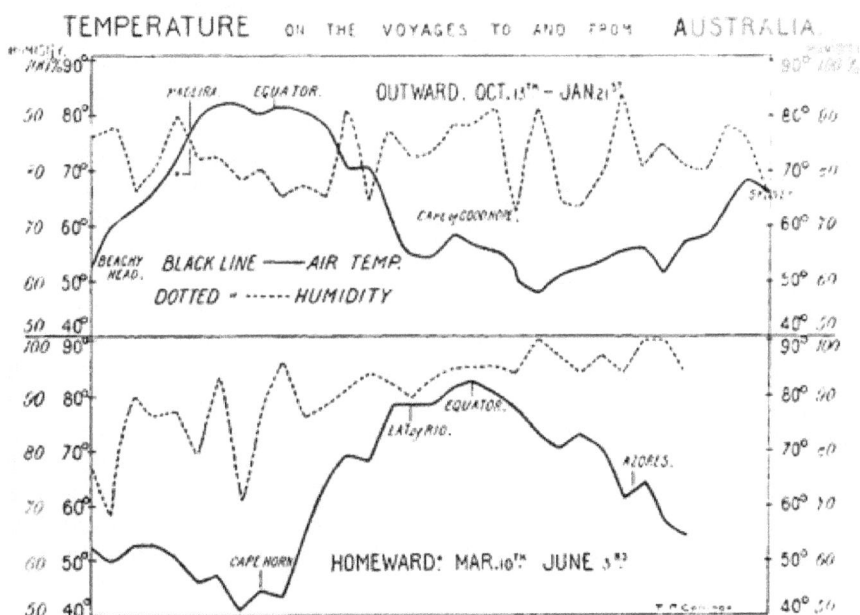

DIAGRAM 5.

ranges from 47° to 55°, and rises to 67° on approaching
the continent of Australia. The actual range of
temperature is from 47° to 82°, the number of rainy
days about 20. The relative humidity varies from
74 to 91 per cent., the average percentage being 82.
When the Australian voyage is made in summer, and
the patient starts in May or June, the temperatures

do not greatly differ from the winter voyage, but the record in passing the equator is 85°, and, of course, Australia is reached in mid-winter.

The return voyage from Australia or New Zealand to England *via* Cape Horn is not so favourable, as the ship starts at 52° or 53°, and often goes southward towards the Antarctic circle, and near Cape Horn the temperature falls to 41°, and for several days is often little over 43°, a rather wretched state of things. In 40 south latitude a rise to 50° occurs, in 35 south latitude to 60°, and off Rio de Janeiro 78° is attained. The line is crossed at about 82°, the Azores at 63°, and England is reached in the early part of June.

The chief consideration in making this voyage is not the season of the year at which it is commenced, but the time of arrival in England or Australia, it being advisable to land in either country in warm weather.

The effect of the first division of the voyage as far as the Cape is sedative, reducing the cough and quieting to the nervous system ; the last, between the Cape and Australia, is a tonic one, when appetite is improved and weight gained.

The voyage to the Cape is a portion of the Australian one, but not so beneficial in its effect, occupying only at present from two to three weeks in a steamer. With two weeks' halt at Cape Town or its environs and the return voyage it ensures an escape of two months of an English winter.

The other winter voyages made by invalids are those to the West Indies in the Royal Mail steamers,

lasting six to eight weeks, though the climate is too
mild for bracing purposes, and the Brazilian voyage.
This latter is the best, as the vessel touches at various
ports—Lisbon, Teneriffe, Pernambuco, Bahia, Rio de
Janeiro, Monte Video, and Buenos Ayres—and
occupies two or three months ; and here experience of
temperate climates is intermingled with that of
tropical.

Very different in its effects on consumptive invalids
is the voyage to India and Australia through the
Mediterranean, the Suez Canal, and Red Sea, for here
we get great heat, and, accordingly, great depression.
During a voyage, leaving England for India the end
of October, the temperature in the Suez Canal rose to
81°, and in the Red Sea to 86·5° F.   At Aden it was
83° to 84°, but after the vessel had left the Straits of
Babel Mandeb it fell to 79°, but throughout the voyage
to Bombay the thermometer records varied from 79°
to 81° F., and after this the course soon became
identical with that round the Cape.   The great
objection to this route is the usually intense heat in the
Red Sea and Suez Canal, where the thermometer
occasionally reaches 98°, which has a most disastrous
effect on patients weakened by chronic disease, and in
many cases induces diarrhœa and great languor and
sweating.   I have known more than one phthisical
patient returning from Australia die in the Red Sea,
overpowered by the results of the great heat, and
many are deteriorated thereby.   The danger on the
homeward trip by this route is the transition, too, from
the heat of the Red Sea and the Suez Canal to the

comparatively cool atmosphere of the Mediterranean,
especially if a tramontana be blowing at the time, a
transition which checks perspiration, and in some
cases, induces temporary albuminuria; as has been
shown by an intelligent surgeon of the P. and O.
Service, who, on testing the urine of the healthy
passengers after entering the Mediterranean from the
Canal, found albumen present in all.

STATISTICS OF EFFECTS OF SEA VOYAGES

I will now furnish some account of 65 phthisical
patients who used sea voyages as a form of climatic
treatment, and see what conclusions may be drawn
from them.

Fifty-seven were males and 8 were females. The
small number of the latter was probably due to
the difficulty and expense of providing escorts or
chaperons. The ages of the males averaged 25·77,
and of the females, 26·5. The length of illness before
starting was on an average 22 months. Family
predisposition was present in 22 cases, or in 33·84 per
cent.; hæmoptysis in 25, or in 38·46 per cent.

Table II. gives at a glance the condition of the lungs
in these patients ; 41, or 63 per cent., were in the first
stage, or that of tuberculisation, 17 having the right
lung affected, 12 the left, and 12 both lungs ; 24
patients, or 37 per cent., were in the softening and
excavation stages ; of these, 5 had a right lung cavity :
5 a right lung cavity with tuberculosis of left lung ;
7 showed a left lung cavity, and 7 a left lung cavity and

tuberculisation of upper right lung. Of the whole
number there was single affection in 63 per cent., and
double in 37 per cent. : 34 per cent. had the right lung

TABLE II.—*Showing the Influence of Sea Voyages on the Condition
of the Lungs in sixty five Cases of Consumption.*

Stage.	State of Lungs before Sea Voyage.	Number.	Percentage.	Arrest of Disease.	Decrease of Disease.	Stationary.	Advance of Disease.	Advance and Extension.	Extension.	Improved.	Stationary.	Worse.
Tuberculisation, 1st stage.	17 had right lung alone affected	41	63	3	6	2	2		4			
	12 had left lung alone affected			2	6	1	1		2			
	12 had both lungs affected				7	1	1		3			
				5	19	4	4		9	58	10	32
Softening and Excavation (2nd and 3rd stages) + Tuberculisation.	5 had right lung in 2nd or third stage	24	37	3	1				1			
	5 had right lung in 3rd stage, and left in 1st stage			4		1						
	7 had left lung in 2nd or 3rd stage			3		2	2					
	7 had left lung in 2nd or 3rd stage, and right in 1st stage			2	2	2	1	1				
				12	3	5	3	1		50	12	38
				5	31	7	9	3	10	55	11	34

	Per Cent.		Per Cent.
Both lungs affected in	24 = 37	Right alone affected in	22 = 34
One lung affected in	41 = 63	Left alone affected in	19 = 29

alone affected, and 29 per cent. had the left lung alone
affected. Summing up, 63 per cent. were in the first

stage; 37 per cent. in the second and third, and 63 per cent. had unilateral disease; 37 per cent. bilateral.

Forty-four made single voyages to their destination and back again, not always returning by the same route, and 21 took more than one voyage, sometimes venturing on a fresh route, sometimes repeating the same, three, or four, or five, or even in one case ten times. The 65 patients took in all 118 voyages, and of these, 48 went to Australia and back round the Cape; 2 to Australia by the Cape, returning by Cape Horn; 2 to Australia returning by America; 2 to Australia to settle; 9 to New Zealand, of whom 6 returned home by the Cape, 1 by Cape Horn, and 2 *viâ* America; 8 to India, round the Cape; 25 to the Cape and back; 8 to the West Indies and back; 1 to China and back; 2 to South America and back; 1 to the United States; 10 sea voyages were made by 1 patient in the Indian Ocean.

It will be seen that the majority took the Australian or New Zealand voyage, and went and returned in clippers round the Cape, with the exception of 9, who either did not return to England or returned *viâ* America or Cape Horn. Consequently we have records of a good number of patients giving a fair trial to this form of sea voyage. The twenty-five voyages to the Cape and back, though much shorter, may be looked upon also as typical examples. The other voyages, though interesting from their effects in individual cases, are too few in number to draw any definite conclusions from as to the influence of each particular route.

The *general* results of the 65 sea voyages are : Cure in 3 cases, great improvement in 21, improvement in 26, making total of improved 50 (or 77 per cent.) ; 1 stationary and 14 (or 22 per cent.) worse. The 41 patients with tuberculisation yielded 78·04 per cent. of general improvement, and the 24 softening and excavation cases, 18, or 43·75 per cent. of improved.

The general improvement consisted of gain of appetite, colour and strength, and above all, of weight; sea voyages surpassing all the other climatic groups in this particular. Tweny-five of these patients were weighed before and after the voyages, and the result was 17 had gained, of whom nine gained more than a stone, and the rest from 5 to 12 lbs. In three cases the fact of gain only was recorded. In 2 patients the weight remained stationary, and 6 lost weight.

The large gain of weight is to be explained by the improvement of appetite, the regularity of the meals, the plentiful supply of food at them, and the lack of opportunities for exercise.

The *local* results show arrest in 5, decrease of disease in 31, stationary condition in 7, advance of disease in 9, advance and extension in 3, extension alone in 10 ; or, joining the cures and decrease, we get a total of improved 36, or 55·38 per cent.: stationary 7, or 11 per cent.; and similarly joining the other categories we obtain a total of worse 22, or 34 per cent. This local result is far less favourable than the general result, as comparison will show, and demonstrates that although patients during sea voyages improve in appearance, and even in the leading symptoms, the disease of the lungs may steadily progress. Hardly more than half improved, and one-third deteriorated. On further examining these cases we find that the tuberculisation cases showed 58·54 per cent. improved and 32 per cent. worse, and the excavation cases showed 50 per cent. of improvement, against 37 per cent. of deterioration. Among the first stage or tuberculisation patients, the unilateral and bilateral improved about equally, the percentages of improvement being respectively 58·62 and 58·34. The large proportion of improvement among the excavation cases, compared with the Riviera and other patients, rather indicates that the sea voyages are more strikingly beneficial in excavation than in tuberculous consolidation, this conclusion being opposed to the opinion held by Dr. Walshe and others. It is possible that a saline atmosphere may promote antiseptic changes within the cavity and thus prevent the spread of local infection and the absence of lung gymnastics and the physiological rest which is a marked feature of life on board ship may promote quiescence of disease and in time fibroid changes.

When we compare the results of the different kinds of sea voyages, we find that the one to Australia comes out best. Of 49 patients 31 generally improved, 1 was stationary, and 9 became worse; or, in other words, 75 per cent. improved and 22 per cent. deteriorated. The local change is not so great, 61 per cent. improving, and 29·27 per cent deteriorating. The Cape cases showed 11 general improvement and 4 worse, which is also a good result against 8 local improvements and 4 worse, or the improvements about double the deteriorations.

An interesting feature, and somewhat opposed to Rattray's conclusions, is that the patients who voyaged to India *via* the Cape improved generally and locally, and the one who took ten voyages in the Indian Ocean improved greatly. All these patients must have been for some time exposed to tropical climates.

With regard to the forms of phthisis specially benefited by sea voyages, I would place the scrofulous or strumous form first ; next the hæmorrhagic type, marked by limited consolidations and large recurrent hæmoptysis ; next, the chronic unilateral cavity cases, without great local irritation. Whether sea air promotes the fibrotic process or no, I cannot say ; but to judge by my cases contraction of cavities takes place very frequently during sea voyages. Cases of early consolidation do not progress so well as in certain other climates, though they do not fare badly.

There is no evidence that sea voyages exercise any special influence on the tuberculous disease, except possibly by promoting fibrosis ; but where there is septic discharge in the form of chronic abscess, fistula, and the like, sea air seems to promote a healing process. The effect is most marked in empyema, where it is often marvellous. I have sent empyematous patients with drainage tubes in their chests on sea voyages, and the result has been highly satisfactory, the discharge gradually ceasing and the cavity granulating up sufficiently to allow of the removal of the tube.

Other cases that profit largely by sea voyages are the mentally overworked, who find in the monotonous routine, amid the soothing influence of a saline atmosphere, that rest which was denied to them in the busy hum of cities. As an example of the benefit of sea voyages I would instance the following three cases :

Case VI.—A warehouse clerk, aged thirty-eight, employed in a dusty atmosphere, but fond of athletics, consulted me on November 1st, 1889, complaining of cough, tinged expectoration, and night sweats, with some, though by no means great, loss of weight. He had a well-developed thorax, but dulness and cavernous sounds were detected in the first interspace on the right side, and crepitation above the left scapula. The sputum contained tubercle bacilli. I prescribed cod-liver oil and a tonic, and advised a voyage round the Cape to Melbourne. He improved slightly under treatment, and started at the end of November. He lost his cough on the outward voyage. On arrival he visited Melbourne and Sydney, and returned home *via* Singapore, Hong-kong, Japan, Vancouver, and the Canadian Pacific Railway, across North America. I examined him in September, 1890, and found him wonderfully better with no increase of weight, but free from cough or expectoration. On the right side there was flattening and dulness with faint tubular sound in the first interspace. No cavernous sounds could be detected on coughing. Crepitation was still audible above the left scapula. He returned to work and, in the following April, 1891, came to me again. He had stood the severe winter well, and gained 11 lbs. No cough or expectoration, but once he had slight hæmoptysis. No signs could be detected in the left lung, and none in the right, save slight flattening and dulness. I hear that he still (1894) keeps well.

Case VII. is that of a law student of Edinburgh, aged twenty-seven, who consulted me on November 9th, 1877. In 1868 he had cough and expectoration, and the late Dr. Warburton Begbie found tuberculous disease in the right lung. He had improved under cod-liver oil and tonics, but the cough persisted, and in the last four months he had lost 9 lbs. He was working hard at his studies. There was no fever or wasting, and I recommended a voyage to Australia. I saw him on his return, in September, 1878, and found him greatly improved. On the outward voyage he had lost his cough and had gained 8 lbs. He remained in Australia three months and a half, and lost 6 lbs. during this stay; returned round Cape Horn, and again gained 4 lbs.; but off the Azores he was chilled by night exposure, and caught cold with fresh cough, which however had subsided when I examined him. I could detect no physical signs in his chest. He returned to his work, and though he remained a good deal in the house during the ensuing winter, he had no return of his symptoms. A careful subsequent examination, in August, 1879, also failed to detect any physical signs.

This case was interesting because the disease, first detected by Dr. Begbie, was of several years' standing when he tried the experiment of a sea voyage, and yet the result was very satisfactory.

Case VIII., an Irish gentleman, aged forty, who consulted me in October, 1877. Five years previously cough had come on, followed by hæmoptysis to the amount of a pint, which was repeated, to a less extent, every two or three months until June, when the hæmorrhage became more frequent, recurring every month during June, July, and August, when it ceased. Each attack as a rule lasted for ten days. Pulse and temperature quiet, and there was considerable yellow expectoration. I detected tubular sounds above the right scapula, and crepitation under the left clavicle and scattered over the whole posterior surface of the left lung. I advised a tonic and cod-liver oil, strong counter-irritation to the left chest, and a voyage to Australia; which last recommendation was not carried out. I saw him in March, 1878, and learnt that he had used counter-irritation largely, and had persevered with the cod-liver oil. He had had several attacks of hæmoptysis, but only in small amounts. The crepitation sounds in the left lung had diminished, and he had gained some weight. He now, on my repeated advice, took a voyage to the Cape, remained twenty-four days on shore, and then returned, and came to me in June, 1878, when I found him well bronzed, without cough or expectoration; breath not short, and the physical examination showed only a small area of crepitation over the left posterior apex; the right lung was clear. He returned to Ireland, and I did not see him again till February, 1886, 7½ years later, when he told me he had been very well till that winter, when his breath was becoming short on exertion, and there was evidence of contraction of the left chest. I came to the conclusion that the left lung was undergoing fibrosis.

This was a well-marked case of hæmorrhagic phthisis, which appears to have become arrested by the voyage to the Cape. Probably fibrosis set in early, though its manifestations were few till some years later. The remarkable feature was the entire cessation of the hæmorrhage from the date of the voyage, though it had recurred with great frequency and regularity previously.

CASES IN WHICH SEA VOYAGES ARE INDICATED

Sea voyages may be recommended in the following cases: (1) Chronic pleurisy and chronic empyema. (2) Chronic bronchitis. (3) Various forms of scrofulous disease, including scrofulous phthisis. (4) Hæmorrhagic phthisis. (5) Tuberculous excavation, where the cavity is limited and the disease unilateral. (6) Neuroses, the result of overwork, and especially insomnia.

Another important group of moist climates are the British South Coast Stations, which were fully discussed in my Lettsomian Lectures,[1] to which those interested are referred.

[1] *Influence of Climate in Pulmonary Consumption.* Smith and Elder.

LECTURE III

of Arrests—Influence of Stage and Single or Double Affection on result—Cases of Arrest—Confirmatory Evidence—Conclusions—Danger of Relapse on Descent—Different Opinion — High Altitude Stations—Alpine—Winter and Summer Climate —Snow-Melting Season—St. Moritz—Davos—Maloja—Wiesen —Arosa—Andermatt—Leysin—South Africa—Dry Climate—Aliwal North—Beaufort West—Matjesfontein in Great Karoo District—Tarkastad, Dordrecht, Burghersdorp and Lemoenfontein in Nieuwveld Mountains—Kimberley—Colesberg—Victoria West—Orange Free State—Bloemfontein—Transvaal —Pretoria—Heidelberg—Johannesberg—Case illustrating good effects of Cape Highland Climate—Rocky Mountain Stations —Advantages—Case of Arrest—The Andes—Cause of Dry Climate—Variety of Climates in small compass—High-lying Plains of Peru, Bolivia, New Granada—Santa Fé de Bogota — Quito—Arequipa—Tarma—Jauja—Huancayo—La Paz—Difficulties—The Himalayan Stations—Landour—Murree—Kasauli —Naini Tal—Simla—Darjeeling—Opinion of Indian Medical Men—Nilgiri Mountains—Ootacamund—Kotaghery—Wellington—Coonoor—The Pulneys—Good Influence of High Altitudes in Imperfect Thoracic Development, Chronic Pneumonia, Pleurisy and Asthma—In what Diseases Contra-indicated— Comparison of Climate results—High Altitudes—Sea Voyages —Riviera and Home Climates—Local and General Results— Importance of Medical Supervision of Climate Patients.

IT is well known that at sea level barometric pressure is about 30 in., varying from a minimum of 27 in. to a maximum of 31 in. ; that it increases on descending into mines, and that, if we could have a shaft sunk vertically forty-five miles into the earth, the air at the bottom would be as dense as quicksilver, while at the summit of the highest mountain of the Himalayas it would be rarefied to the extent of two-thirds. But what we have to deal with is the influence of lower degrees of density or of rarefaction on the human body.

INCREASE OF PRESSURE

Taking the subject of increase of pressure first, we find our knowledge has been chiefly derived from the experiments of diving bells, pneumatic tubes, or caissons, and the effects on divers.

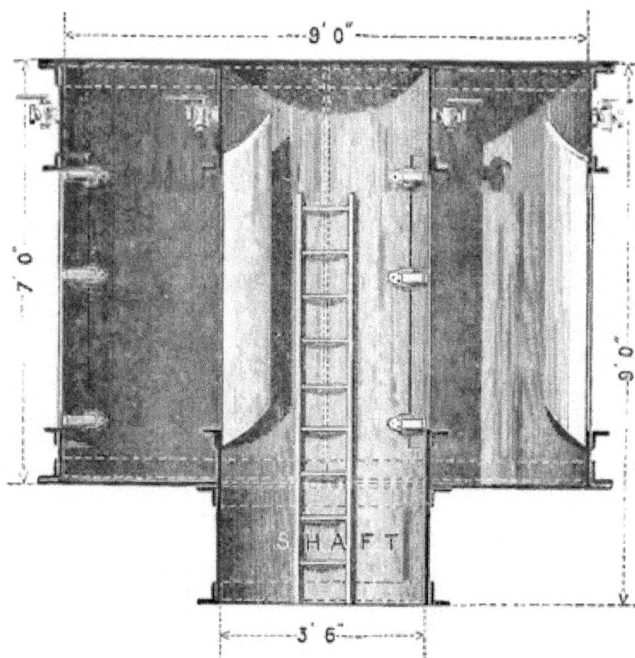

DIAGRAM 6.—AIR LOCK OF FORTH BRIDGE; VERTICAL SECTION.

The symptoms on descending in diving bells depend on the depth of water to be traversed, as, of course, the compression of air increases with the depth; but at about 30 ft. deep, pain and noises in the ears are felt, with a sensation of the head being bound with

iron. According to Colladon, these cease on the
bottom being reached, and the ascent is not disagreeable, except for a feeling as if the bones of the skull
were separating. No change in pulse or respiration
is noted, and the above-mentioned symptoms seem due
to the rapid descent rather than to the compression of
air. Owing to there being no ventilation of these

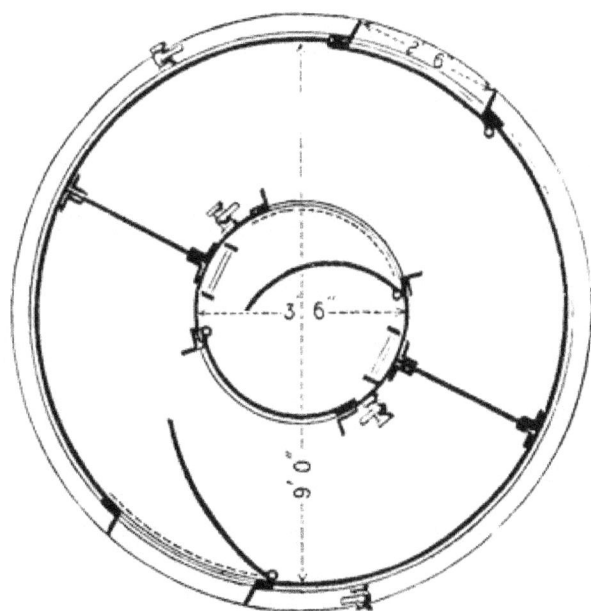

DIAGRAM 7.—AIR LOCK OF FORTH BRIDGE ; TRANSVERSE SECTION.

bells, they were soon replaced by diving dresses, to
which air was freely supplied, and by pneumatic tubes
where atmospheric pressure was used to keep back the
flow of water. In these tubes air is compressed to the
extent of three to four, and even more, atmospheres,
and a large number of labourers remain at work for
several hours at a time.

Caissons.—It may be as well to explain the means employed in pneumatic tubes to accustom the workmen to the great contrast of pressures between the tube and the external air. The caisson is connected with the upper world by shafts which the workmen descend, but at the top or bottom of the shaft—the former to be preferred—is an air lock or iron ante-chamber, well shown in the annexed drawing of the Forth Bridge air lock, furnished by Mr. Moir, one of the engineers, which the men enter, and, closing an air-tight door behind them, admit the compressed air from the caisson by means of a turncock till the pressure has reached the same degree as in the caisson itself, when a communicating door is opened and the men enter. Similarly on leaving the caisson or shaft they shut off the lock and reduce pressure to the normal by blowing off air through a tap into the external atmosphere. It is manifest that on the management of this lock or chamber depends the safety and even life of the workmen, and the increase or diminution of pressure should always be made gradually and at a fixed rate. Mr. Moir found that they could endure a rate of pressure diminution of 5 lbs. a minute without danger or suffering, and that most of the accidents occurred when this was exceeded.

CAISSON DISEASE

Long exposure to this atmosphere, and more especially the change to air at ordinary pressure, gives rise to various symptoms which have been well

described by Pol and other observers, and lately
collected together under the title of " caisson disease,"
by Dr. Andrew Smith,[1] of New York.

The symptoms are :—severe and often excruciating
pains coming on suddenly, commencing in or near
both knees, and extending upwards into the trunk,
associated with gastric pain and vomiting. The pain
principally affects the muscles. In some cases it is
followed by paralysis, generally paraplegia of lower
extremities, and paralysis of the bladder and rectum.
Cerebral symptoms, such as vertigo and headache, are
frequently present. The symptoms appear to be the
result of congestion of the brain and spinal cord,
resulting in sanguineous effusion and congestion of
most of the abdominal viscera. The pulse is rapid at
first—that is, on leaving the tube—but quiets down
afterwards, and resumes the normal in half an hour.
The skin is cool, perspiring, and pallid, often of a
leaden tint. The temperature is normal, paralysis
is present in various degrees. In Dr. Jaminet's St.
Louis cases it occurred in forty-seven out of seventy-
seven workers, or 61 per cent. ; at New York, Dr.
Andrew Smith found it only in 15 per cent., the
difference being explained by the higher pressure used
in the St. Louis caisson.

Paralysis most frequently affects the lower half of
the body, but it may include the trunk or one or both
arms, and affects sensation as well as motion, but the
pains in the limbs may continue when they have

[1] *The Physiological, Pathological, and Therapeutical Effects of
Compressed Air.*

become insensible to pinching or to pricking. The
duration of caisson disease varies from three or four
hours to six or eight days. The neuralgic pains do
not generally last more than twelve hours, but have
sometimes extended to five or six days. The paralysis
often disappears in twelve hours, but sometimes lasts
weeks. Death usually takes place from coma. The
necropsies of patients dying from this disease show,
according to Dr. van Rensselaer, congestion and softening of the spinal cord and congestion of the brain. In
a few cases meningitis, both cerebral and spinal, was
present, and in one case effusion of blood between
the arachnoid and the pia mater, but, as most patients
recover, the necropsies are few.

The symptoms of caisson disease never appear in
the tube itself, but after leaving it, and depend partly
on the time spent in the tube, but chiefly on the
rapidity with which pressure is diminished, which
never ought to exceed 5 lbs. a minute. Some workmen appear more easily affected than others, and the
predisposing causes to attack seem to be :

1. Inexperience of the work, newcomers being more
easily affected than old hands.

2. Fulness of habit, the full-blooded heavy-built
man being more frequently attacked than the slim and
wiry.

3. Severe exertion after leaving the caisson, such
as ascending a long staircase, which Dr. Smith says
may be avoided by using elevators, or by placing
the air lock at the top of the shaft as in the Forth
Bridge, instead of at the bottom, as all muscular

exertion is more easily borne in the condensed air
than in air at normal pressures.

4. Fasting before entering the tube. The remedy
in the event of the caisson malady appearing in the
air lock is to increase the pressure, when the symptoms
gradually disappear.

DIVERS' DISEASE

Divers use compressed air in their operations at a
depth of 54 metres in the sea, but they have to deal,
in addition, with the pressure of the water on their
bodies, which at that depth equals 6·4 atmospheres.
They suffer from the same symptoms as the workers
in pneumatic tubes, only more severely, and it is
calculated that the Greek sponge divers have a
mortality of 10 per cent. from this cause alone. They
have prickings ("*les puces*," as the French call them),
muscular pains which chiefly affect the muscles most
used by the divers; pains in the joints, paralysis of
different parts, and often paraplegia, including paralysis
of the bladder and sphincter ani; all these symptoms
generally supervening shortly after the diver has left
the water. A *post-mortem* examination made in one
case showed extravasation of blood between the spinal
dura mater and the arachnoid, and the greater part of
the cord in a state of softening.

THERAPEUTIC USES OF COMPRESSED AIR

We now turn to the therapeutic uses of compressed air, for the application of which two sets of apparatus have been devised.

Methods by which Air is Inspired through a Mask tightly fitting to the Mouth.—A hollow metal cylinder closed at one end is plunged into a second or inverted cylinder containing water, in which it floats owing to the contained air. By means of pulleys and weights, an equilibrium is established, and a pipe is passed from the air cylinder through a drying box to the mask fitted to the patient's nose and mouth, enabling him to respire the air, which can be either rarefied by drawing off the water, or condensed by placing weights on the top of the air cylinder. Hauke's was one of the earliest forms of this ; Waldenburg's, which is better known, is a modification of it. Some apparatus, like Cube's and Schnitzler's, consist of two cylinders, one for condensing, and the other for rarefying, which can be alternately connected with the patient's mouth ; others are constructed on the principle of the centrifugal pump, by which air is stored up by the action of water. The forms of such apparatus are very numerous, and are to be found elsewhere,[1] but all are open to two objections : (1) that the use of the mask is exceedingly irksome, and often induces headache and faintness ; and (2) that it is impossible to keep up a proper supply of air at the

[1] Oertel, *Respiratorische Therapie.*

requisite pressure, and that consequently there is a
danger of re-breathed air.

Compressed Air Baths.—The second form of
applying compressed air is by surrounding the patient
with an atmosphere of condensed air, as in a bath,
a form largely used on the Continent, where no fewer
than fifty establishments for this purpose are in operation. The subjoined drawing (Diagram 8) of the
Brompton Hospital compressed air bath, manufactured
by Haden, of Trowbridge, will explain the essential
elements of the apparatus. For a compressed air bath
are necessary :—

1. A strong circular or ovoid wrought iron chamber
(c) with arched roof, with walls not thinner than three-
sixteenths of an inch, and strengthened with girders
and ribs of iron and provided with thick glass windows
and a stout closely-fitting door. The size of the
chamber to be regulated by the number of persons
to be accommodated. The circular one at the Brompton Hospital is arranged for four, and measures
10 feet in diameter by 8 in height. The chamber
must be furnished with inlet (w) and outlet (w) pipes,
and an air-tight cupboard to pass in food and messages.

2. A compressing apparatus (A), steam or water
power being usually the motor.

3. A reservoir (B) for the reception of air, and to
purify or cool it previous to use.

4. Valves to regulate the current to the chamber.
The air is supplied from a pure source, and then
filtered through cotton wool and, as it were, pumped
through the bath, as accumulation is attained by the

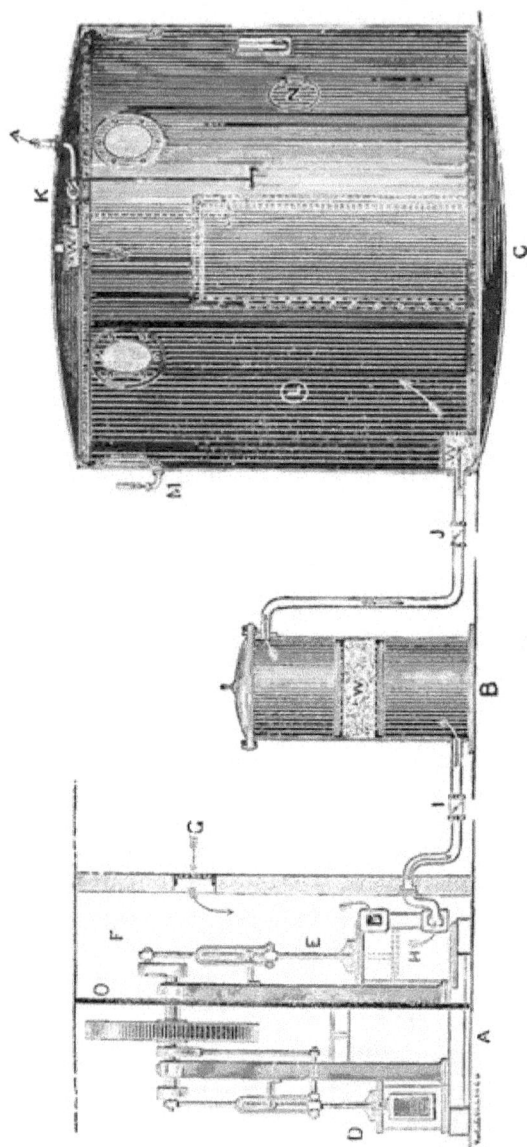

DIAGRAM 8.—COMPRESSED AIR BATH APPARATUS.

D, steam engine which by means of a flywheel and crank works a second engine E, in air compartment. Air from outside enters chamber through G aperture, following line of arrows, passes into cylinder, and penetrates piston plate H, which is perforated by diaphragm valves (not shown) closing during descent and opening during ascent. It is then driven into B, receiver, and C, air-chamber, the valves I and J preventing reflex.

outlet being smaller than the inlet, and thus many changes of atmosphere are obtained during the sitting, which lasts, as a rule, two hours, half an hour being spent in increasing pressure, one hour in maintaining it at the highest point required, and half an hour for reducing it.

The pressure, for therapeutic purposes, seldom exceeds 10 lbs., two-thirds of an atmosphere, and the increase and decrease is therefore made at the rate of 1 lb. in three minutes. Air invariably rises in temperature during compression and cools during rarefaction, and in summer it is sometimes a great difficulty to keep the chamber cool even with ice packing in the reservoir.

The effect of compressed air at this moderate pressure on healthy individuals differs considerably from the effect at the high pressures of 30 lbs. and 40 lbs. used in pneumatic tubes. The first sensation, as pressure is increased, is an unpleasant one in the throat, referred to the pharynx immediately behind the tonsil, which is relieved by swallowing saliva or drinking water. Pain is also felt in the membrana tympani. These sensations are due to the different calibres of the external auditory meatus and the Eustachian tube. The Eustachian tube being very much smaller than the external auditory meatus, the column of air penetrates with difficulty to the internal surface of the membrana tympani, and changes of pressure are slowly communicated, whereas through the meatus air passes freely, and causes, under these circumstances, a convexity inwards of the auditory membrane. This is owing to increase of pressure ; the opposite takes place during reduction of pressure, for the middle ear is full of air of greater density as the change in it is slow, whereas the external meatus contains air of which the pressure is reduced rapidly ; this gives rise to convexity outwards of the membrana tympani, hence the unpleasant sensations of the ear and throat at the beginning and end of the bath. It is said that all the special senses are impaired, but I cannot say I have found this. The voice becomes shrill, and singers often gain a note or two higher than their average in the bath.

Some years ago I placed two remarkably healthy well-made house-physicians of the Brompton Hospital in the compressed air bath, where they remained three hours and a quarter, and most of my conclusions as to the effect of compressed air on normal subjects are the results of observations carefully taken by these gentlemen on each other, and afterwards checked by myself. The effect on respiration of the compressed air bath is that the individual finds he breathes slower, deeper, and more easily. The respiration rate, according to my observations, falls from 16 or 20 to 14 or 15 at least. Von Vivenot found it fall to 4 or even 3 a minute. Inspiration becomes very easy, but expiration is less easy ; the ratio between them undergoes considerable modification, expiration being sometimes twice or three times as long as inspiration. The increased depth of the respiration is shown by Lowne's spirometer, which invariably gives an increase in the amount of air expired.

It would appear that breathing compressed air in creases lung capacity, probably by opening up more alveoli, which had previously not been brought into use, and we must suppose that the diminished number of respirations means that their amplitude makes up for their smaller number. The effect on the circulation is that the pulse is slower, smaller in volume, but of increased arterial tension, the capillaries smaller, and the veins less full of blood. The pulse rate diminishes 4 to 20 beats a minute, but on returning to the outer air it returns at once to the normal. Sphygmographic tracings show a lowering in the height of the tidal and dicrotic waves.

The effect of the pressure on the circulation was admirably shown by Von Vivenot's observations on a white rabbit in the bath. Under normal pressure, with the rabbit quiet and at liberty, the ears were full of blood, the conjunctival vessels injected, and the iris tinted deep red, but in a compressed air bath the vessels of the conjunctiva became finer and more pale, and in one experiment they alternately filled and emptied. When pressure was maintained at the maximum the iris and pupil became decolorised, and the ears, seen by transmitted light, showed empty vessels, and the larger vessels were scarcely visible.

From these latter observations the conclusion is that compressed air exercises an intropulsive influence on the circulation, affecting those surfaces most exposed to it, such as the skin and lungs. The blood is thus drawn into the organs protected from air pressure, namely, the brain, heart, liver, kidneys, and spleen.

The pressure is exerted more on the capillaries and superficial veins than on the deeper veins and arteries. and its tendency would be to reduce pressure on the right side of the heart,[1] and to increase it on the left. Dr. Burdon Sanderson thus accounts for the slower pulse rate, " the effect of the diminished fulness of the venous system is to retard the filling of the ventricles during the period of relaxation, and consequently to lengthen the diastolic period, and thus diminish the frequency of the pulse."

The introduction of a larger amount of oxygen causes greater absorption by the lungs, and leads to further oxidation and tissue change ; this being proved by the bright colour of the blood, seen during bleedings in the bath, by the increase of the carbonic acid exhaled from the lungs, and of urea excreted by the kidneys. Muscular power is augmented, appetite generally improves, and weight is almost invariably gained. The temperature is not materially affected.

EFFECTS OF COMPRESSED AIR IN LUNG DISEASES

We will now consider the effects of compressed air on various lung diseases, and I may state that my conclusions in the present instance are based on cases from my private notebooks and on 66 patients suffering from various forms of lung disease under my care at the Brompton Hospital treated in these baths,[2] most

[1] *Practitioner*, October 1868.

[2] Further details of this treatment are given in my lectures on the *Compressed Air Bath and its Uses in the Treatment of Disease*. London : Smith, Elder, and Co. 1885.

carefully tabulated by my late house-physician, Dr. Horrocks.

Asthma.—Of bronchial asthma there were 15 cases —10 males and 5 females. In 7 asthma was largely complicated with emphysema. The average number of baths taken was from 12 to 15. Of these patients 12 improved and 3 did not improve. Out of 11 whose weights were taken, 9 gained weight—on an average 4⅔ lbs.—and 2 lost. The measurement of the circumference of the thorax at various levels was made in 7 patients, before and after the baths, and the circumference increased in 4 and diminished in 3. The spirometer showed an increase of 25 to 33 per cent. in the 3 cases in which this test was applied. The principal effect of the baths on asthma appears to be sedative to the pulmonary plexuses of nerves and to the pneumogastric. The attacks are rendered less severe, and after a course of twenty or thirty baths, the intervals between the attacks become much longer. I do not remember one case where a complete cure was effected, but I recollect several where the patient remained free for months, and, in one instance, for years, from asthma. The effect on the paroxysm is immediate and wonderfully efficacious, in fact more so than any medicines, and many asthmatics have expressed to me the wish that they could live in the bath and thus be freed from their sufferings. The effect on the emphysema accompanying it, is to reduce it, as percussion and auscultation show.

Chronic Bronchitis and Emphysema.—In chronic bronchitis and emphysema the effect is satisfactory:

the cough diminishes and the expectoration is lessened; weight is gained; breathing is easier; but the great feature is the reduction of the emphysema. Examination of the chest shows diminution of the hyperresonance and a return of the various displaced organs to their normal positions, and cyrtometric measurements give a reduction of the chest circumference at different levels, a reduction varying from $\frac{1}{2}$ to $1\frac{1}{2}$ inches. This indicates that a great deal of emphysema, even in adults, is of a temporary nature, produced often by severe paroxysms of coughing or dyspnœa, and capable of reduction, if respiration be rendered more easy, as in a compressed air atmosphere. In my 33 cases of chronic bronchitis and emphysema, who had on an average eighteen baths, 15 were measured cyrtometrically, and of these 11 were found to have decreased in circumference and 4 to have increased; 26 of the patients improved generally, 5 did not improve, and 2 died, one of heart failure from cardiac dilatation after five baths, and the other from an attack of capillary bronchitis, after having greatly improved from nine baths, when he contracted fresh bronchitis and died later.

Phthisis.—My experience of compressed air in phthisis is not altogether favourable. In 9 of the cases I submitted to the bath there was gain of weight and some diminution of cough and expectoration, and apparently the respiration became freer in the unaffected portions of the lungs, but in two cases the bath appeared to bring on hæmoptysis, and in 4 patients hæmoptysis came on during the treatment,

though it could not be distinctly connected with it.
Beyond the opening up or aeration of portions of the
lung which had not been brought into play for some
time, I could see none of the improvement resulting
from compressed air which is so loudly proclaimed by
Oertel and Simonoff, nor could I discover that it
facilitated the absorption of lung consolidations or infiltration, though cases of this class were submitted to
the bath ; or, lastly, that it promoted, as Oertel states,
the absorption of serous exudation in acute pleurisy,
and tended to expand the compressed lung.   My experience is that it exercises no effect in expanding a
lung which has been compressed with fluid when the
fluid has been removed, and, even after a course of
air baths steadily persevered in, the fluid will reaccumu-
late, and will make itself known by indubitable physical
signs, and by the diminishing amount of expiratory
power as evidenced by the spirometer.

DECREASE OF PRESSURE

The influence of diminished atmospheric pressure,
or a rarefied atmosphere, on the human body has been
tested in more than one way.   Aëronauts have ascended
in balloons to great heights, and Messrs. Glaisher and
Coxwell attained an elevation, where the barometer
showed only 9¾ in. pressure, equivalent to 29,000 feet.
The balloon rose still higher, but Glaisher had lost
consciousness and could no longer register, and so the
higher elevation was never recorded, though Coxwell
fancied he saw the barometer stand at 7 inches, equi-

valent to a height of 37,000 feet. Mountain climbers.
like Mr. Whymper. have reached elevations of 22,000
feet. and have borne the rarefaction. sometimes suffering
from the soroche or mountain sickness, sometimes
entirely escaping. the symptoms not always coming on at
the greatest height, but in Mr. Whymper's case at 16,000
feet.   In South America, after acclimatisation, large
populations live at great elevations, and both there
and in North America mining communities are to be
found at altitudes up to 10,000 or 11,000 feet, and in
the Rockies the formation of the country rather lends
itself to residence at high levels, as there is a gradual
rise from the valley of the Mississippi to the west, and
step by step till the Rockies themselves are reached :
and thus we have a vast tract of country lying at the
same altitude as the high Swiss valleys of the Engadine
and of Davos. in a more southerly latitude, and with
a warmer and drier climate, containing towns of
considerable size.

CLIMATE OF HIGH ALTITUDE SANITARIA

The climate of high altitude sanitaria necessarily
varies with height, and latitude, and position in reference to shelter, winds, and rainfall ; but they all
possess, in addition to rarefaction of atmosphere, the
quality of diathermancy, or the increased facility by
which the sun's rays are transmitted through the attenuated air.   The last characteristic arises from rarefaction ; and, according to Dr. Denison. causes an
increase in the difference between sun and shade tem-

peratures of 1° F. for every rise of 235 feet. Hann noted that in the plains 30 to 40 per cent. of the total amount of the sun's heat was absorbed by the atmosphere, whereas at the summit of Mont Blanc. 15,730 feet, the quantity absorbed was not more than 6 per cent. High altitudes are stated to be aseptic, and this is true as long as there is no special cause to pollute the air. Both in the Alps and Rocky Mountains infusions of meat will keep for a long time exposed to the air in winter, but in the mountain villages, where there are large aggregations of human beings or of cattle and horses. the air is no longer purely aseptic, and septic organisms find their way into the infusions. causing putrefaction. Perhaps the diathermancy is the most striking feature of mountain climates, as it affords an explanation of the great solar temperatures which prevail during the day at high altitudes, and of the extreme nocturnal radiation.

The physiological effects of mountain climates on the organs and functions of patients and of healthy persons are very striking. and have been described by me elsewhere.[1] They are briefly as follows. The skin is tanned by the solar rays, and more especially, according to Dr. Bowles, by the ultra-violet rays ; the circulation is at first quickened, and the heart's impulse becomes more powerful, but at the end of six or eight weeks the pulse rate becomes less rapid, and it eventually is slower than the normal (Ruedi). The respiration is at first quickened, but after six or eight weeks it also is found to have slowed, and to have gradually

[1] *Trans. International Medical Congress*, 1881.

fallen to a rate below the normal. The breathing becomes deeper, the inspiration longer, and the expiration more complete (Ruedi).

The quickening of the circulation and respiration before acclimatisation is often accompanied by great thirst and by a reduction in the blood pressure and in the amount of urea excreted by the kidneys ; but on the other hand more carbonic acid (Marcet) and water are eliminated by the lungs. When acclimatisation is complete, the urea appears in the urine in full quantity and the blood pressure increases.

Coincidently with the slowing of the pulse and respiration there is an extension of the thorax in various directions, causing increase in circumference of 1 to 3 inches at different chest levels. There is also an increase of mobility of the thoracic walls. The cause of this is to be found in the greater physiological activity of the pulmonary organs, due to rarefaction of the atmosphere and consequently their more complete development, and this conclusion is confirmed by two facts : first, that this enlargement is most marked in those who take much exercise at high altitudes ; secondly, that it is not always permanent, the thorax and its contents sometimes, though not invariably, reverting to the normal size after return to residence at sea level. Appetite increases, and in some cases weight is gained, though this is not invariably the case, but both muscular and nervous vigour are augmented.

More oxygen is absorbed by the blood and more carbonic acid is exhaled. Paul Bert found that the blood of a llama of Peru living at an altitude of 4,000 metres consumes 18 to 20 per cent. of its volume of oxygen, whereas in Paris similar animals consume only 10 to 12 per cent. of their volume.

Another point established by Viault and Egger is the somewhat rapid increase of red corpuscles, which counteracts the possible danger from anæmia, which Jourdanet and others have insisted on.

The above effects of mountain climates on the low-lander are confirmed by the features of mountain races themselves. The Indians of the Andes, the guides of the Alps, the chamois hunters of the Tyrol, the hill tribes of the Himalayas, all present the characteristics of a large thorax, deep inspiration, and great power of endurance for walking. Mountain races are, on the whole, wonderfully immune from disease, especially from scrofulous and phthisical disease, but this immunity depends on their avoidance of overcrowding, indoor life, and insufficient dietary, for in the presence of these no altitude could give immunity.

EFFECTS OF HIGH ALTITUDES ON PHTHISIS

The use of high altitudes in the treatment of phthisis was, according to Dr. Archibald Smith, an established practice in South America for many years before it was introduced into Europe to the notice of medicine by Drs. Archibald Smith himself, Guilbert, Lombard, Jourdanet, Brehmer, and others, but this country owes its first clear scientific knowledge of the treatment to Dr. Hermann Weber, who contributed a valuable paper to vol. lii of the *Medico-Chirurgical Transactions*,

entitled " On the Treatment of Phthisis by Prolonged
Residence in Elevated Regions." which first directed
my attention. and I believe that of others. to the
subject. and caused me to pay a special visit in 1872 to
Davos and St. Moritz with the object of investigating
the climate and its effects. What I found by examining German consumptives at Davos—for there were
then no English there—convinced me of the value of
the treatment, and as soon as proper arrangements
were made for the reception of English invalids, in
the shape of an English-speaking doctor and hotels
that would cater for English and not only for German
requirements, I began sending patients, and I believe
I have sent over 300 altogether to winter in these and
other high altitude sanitaria. Of these, 247 cases of
phthisis have been tabulated and form the basis of
statistics which will now be given.

Sex.—183 were males, 64 were females. or. roughly
speaking, three-fourths of the former and one-fourth
of the latter.

Ages of High Altitude Patients.

	Males.		Females.		Total.	
	No.	Per Centage.	No.	Per Centage.	No.	Per Centage.
10 to 20 years	21	11·00	14	21·87	35	14·16
20 to 30 ,,	95	51·91	36	56·25	131	53·03
30 to 40 ,,	48	26·30	11	26·56	59	23·88
40 to 50 ,,	19	10·38	3	4·70	22	8·90
Total	183	—	64	—	247	—

Ages.—The average age for the males was 28 years,
and for the females 25·5. As will be seen in the

adjoining table, the great mass of both sexes were between 20 and 30 ; 11 per cent. of the males and 22 per cent. of the females were under 20, and only 7 per cent. of both sexes over 40.

The length of illness before climatic treatment was for the whole number 23·89 months ; for the males, 19·06 months ; for the females, 35·04 months ; and this average was thus composed :—102 had been ill for less than a year, some for only 2 or 3 months, others for 2, 3, 4, 5, 6, 7, or 8 years. 1 had a history of 9 years, 1 of 10, 2 of 16, and 1 of 20 years.

Family predisposition was present in 97 cases, or 40 per cent. ; hæmoptysis in 112, or 45 per cent.

History and Nature of Cases.—The following complications existed : 5 of these patients had malformations of the thorax, of the pigeon-breast type, 1 had spinal curvature, 4 strumous abscesses of various parts of the body, 1 fistula *in ano*, 2 suffered from tuberculous testicle, 5 had syphilis, 4 had cardiac lesions (1, aortic regurgitant and 3, mitral regurgitant disease), 1 had laryngeal catarrh (not tuberculous), 3 had dry basic pleurisy, 11 were cases of hæmorrhagic phthisis, and 2 of the catarrhal form ; the rest of the patients were examples of chronic phthisis, as a rule without pyrexia, the existence of which would at once have precluded the recommendation of high altitudes, my experience being that the climate generally augments pyrexia.

State of Lungs.—Table III. shows the state of lungs of these 247 patients before starting for high altitudes, and from this it will be seen that 161, or 65 per cent.,

were in the first or tuberculous stage, 67 having the
right lung, 42 the left lung, and 52 both lungs
affected.　The extent of tuberculosis varied consider-

TABLE III.—*Showing the Influence of High Altitude Climates on the
Condition of the Lungs in 247 Cases of Consumption.*

Stage	Number	Percentage	State of Lungs before Residence at High Altitudes.	Arrest of Disease.	Partial Arrest.	Decrease of Disease.	Stationary.	Advance of Disease.	Advance and Extension.	Extension.	Unknown.	Arrest.	Improved.	Stationary.	Worse.
												%	%	%	%
Tuberculi-sation, 1st	161	65	67 had the right lung alone affected.........	44	15	3	1	1	1	2					
			42 had the left lung alone affected.........	27	8	1	—	1	2	1	2				
			52 had both lungs affected	20	18	2	3	2	5	2					
				91	41	6	4	4	8	5	2	57	87	2	11
Soften-ing and Excavation + Tuber-culisation	86	35	25 had right lung in 2nd or 3rd stage............	6	8	3	1	3	2	2					
			12 had right lung in 2nd or 3rd stage, and left in 1st stage............	1	3	2	2	—	2	2					
			22 had left lung in 2nd or 3rd stage............	4	7	1	3	4	2	1					
			26 had left lung in 2nd or 3rd stage, and right in 1st stage ...	3	7	2	3	2	6	3					
			1 had both lungs in 2nd stage	—	—	—	—	1	—	—					
				14	25	8	9	10	12	8		16	55	10	35
			Totals.........	105	66	14	13	14	20	13	2	43	76	5	19

Per Cent.
Both Lungs affected in 91 = 37
One lung affected in 156 = 63

Per Cent.
Right alone affected in 92 = 37
Left alone affected in 64 = 26

ably, from disease of an apex to that of a lobe, but in
all cases the lesion was undoubted, and in many the

consolidation extensive. Softening and excavation
were present in 86, or 35 per cent. ; in 37 patients in
the right lung, in 48 in the left lung, and in 1 in both
lungs. Of the right lung cavity cases the disease was
limited to the right lung in 25, in 12 the left lung was
also tuberculised. Of the left lung cavity cases the
disease was limited to the left lung in 22, and in 26
the right lung also was affected. Summing up, 65 per
cent. of the patients were in the tuberculisation stage,
and 35 per cent. in the softening and excavation stage.
The disease was unilateral in 63 per cent., and
bilateral in 37 per cent. ; the right lung alone was
affected in 37 per cent., and the left alone in 26 per
cent.

Length of Residence at High Altitudes.—The
average residence of these 247 patients at high altitudes was 12·22 months, a little above a year, but
this large average is deceptive, as it includes a certain
number of persons in whom the disease has been
arrested, but who have become residents in the sanitaria. The great mass of these patients have been,
on an average, ten months, or about two winters each,
at high altitudes ; 123 patients passed collectively 128
years and 8 months at Davos, 1 spent six months at
Arosa, 108 spent 72 years 11 months at St. Moritz,
10 passed 4 years 4 months at Maloja (this was when
the hotel was open in winter), 2 passed 13 months
at Wiesen, 9 passed 22 years 10 months on the South
African Highlands, and 6 spent 119 months in Colorado
and New Mexico.

General Results.—Among the 246 patients a cure

was effected in 101, or 40·89 per cent. ; great improvement in 73, or 29·55 per cent. ; improvement in 32, or 12·95 per cent. ; making a total of improved 206, or 83·40 per cent. ; 5 patients remained in a stationary condition, and 36, or 14·57 per cent., became worse. Roughly speaking, nearly seven-eighths improved and one-eighth deteriorated, of whom 48, or 19·43 per cent., died.

Local Results.—I have somewhat varied the classification of results from that of the other groups, making two classes where formerly only one existed, namely, arrest and partial arrest. By arrest is meant in the case of first stage patients, disappearance of all physical signs of disease. This is sometimes so complete that the medical examiner has to refer to his notes to ascertain which lung was affected. In the case of softening and excavation cases, arrest signifies disappearance of cavernous sounds, and even of signs of consolidation, and nothing to be detected beyond deficiency of expansion, some harsh breathing over the whole side, and tubular breathing or prolonged expiration above the scapula. In some cases not even these signs are present. In contracted cavity cases there was a certain degree of immobility of the side and flattening of the thoracic wall, combined with hyper-resonance and prolonged expiratory sound.

Partial arrest means that though there is evidence of arrest of the disease, some physical signs remain, indicating limited consolidation or a contracting cavity.

Arrest took place in 105, or 42·85 per cent.

Partial arrest in 66, or 26·83 per cent.

Decrease took place in 14, or 5·71 per cent.

This makes a total of improved of 185, or 75·51 per cent.

In 14 there was advance of disease ; in 20 advance and extension ; in 13 extension alone, making a total of 47 worse, or 19·18 per cent. ; and in 13 patients, or in 5½ per cent., the disease remained stationary.

Among the 159 first stagers, there were 91, or 57·23 per cent. of arrests ; 41, or 25·78 per cent. of partial arrest, and 6 decreases ; making the large total of 138 improved, or 87 per cent. ; and only 10·69 per cent. worse. The improved among unilateral cases mounted up to 91·5 per cent., the right lung being a shade more favourable than the left lung, but where both apices were involved, the percentage of improved was 76·5.

The most surprising feature is the large figure of absolute arrests, which not only were far more numerous than in other climatic results, but were more complete. The "decrease of disease" only forms a very small item in the list.

Now we come to the second and third stages, and we see that in 14, or in 16 per cent., there was "arrest of disease" in 25, or 29 per cent. "partial arrest," and in 8 "decrease of disease," or making a total improved of 47, or 54·5 per cent. Nine patients remained in a stationary condition, and 30 showed either advance, advance and extension, or extension of disease, making a total of 30, or 35 per cent. of worse. The rise in the proportion of the worse among these, as compared with those in the first stage, is very striking, the number being doubled ; and this is a good deal due to the "double affection" patients ; as the percentage of improved among the unilateral cavity cases is for the right, 68 per cent., and for the left, 54½ per cent. ; whereas for the bilateral it is 49 per cent.

Nevertheless the prognosis of a cavity case in mountain climates is very inferior to that of a limited tuberculisation case, the ratio of benefit being as 61 to 91, or nearly 2 to 3, and apparently double affection of the first stage improved more than limited cavity cases in the proportion of 76·5 to 61.

The evidence seems to be all in favour of sending cases of tuberculisation, even when both lungs are affected, to the mountains, and the earlier the better.

A few illustrative cases of the influence of mountain treatment may now be given.

Case IX.—Mr. J. L. B., *æt.* 19, medical student, consulted me, Oct. 21, 1886, with a history of cough and expectoration and some loss of flesh of 6 weeks' duration. Sputum contained tubercle bacilli. *Slight dulness, prolonged expiration and bronchophony over first intercostal space on right side;* bronchophony being the most marked of the physical signs. Recommended his wintering at Davos.

May 4th, 1887.—Wintered at Davos, skating, walking, and tobogganing, and lost cough. Was remarkably well whole season, and gained one stone in weight, and chest developed considerably in breadth and depth. *No dulness, but hyper-resonance of chest everywhere. No bronchophony. Still prolonged expiration over first interspace of right chest*

No cough or expectoration. He returned to medical studies, and remained well till the spring of 1891 when he overworked for scholarships, and came to me in the autumn with cough and expectoration, which contained no tubercle bacilli, and with some loss of flesh and strength. After a holiday he soon recovered, lost his cough, and in the following year gained almost all the prizes and scholarships possible at his hospital and remains (1893) in excellent health.

Case X.—Miss C., aged 18, had lost a sister from phthisis and was seen by me July 20th, 1887, with a history of cough and expectoration for 5 months, and wasting and night sweats for 2 months ; total loss of appetite ; aspect very pallid. *Slight dulness, crepitation in first interspace on right side.* Ordered to St. Moritz for the winter.

May 17th, 1888.—Wintered at St. Moritz, spent 6 weeks of the spring at Wiesen. Entirely lost cough and expectoration, gained 1 stone 10 lbs. in weight, and became well bronzed, looking the picture of health. Her chest has increased enormously in circumference, and measures, on full expiration, 5 inches more at the level of the second rib than before she left England. States that she has burst all her clothes. Careful examination shows great development of the thorax and hyper-resonance everywhere, but no abnormal physical signs.

September, 1891.—Has been very well till the last few weeks, when became engaged to be married and very anxious. Has lost flesh, and chest circumference is not so large as formerly. No cough or expectoration, and though thin she appears in good health.

Here is an interesting case where other climates were tried first :

Case XI.—Miss R., *æt.* 21, was sent to me by Dr. Owen, of Beaumaris, November 22nd, 1879, with a history of cough with expectoration, loss of flesh, night sweats, pain in the left chest and evening pyrexia of a month's duration. *Dulness and deficient breath sound were detected close to the left scapula.* She was put on cod-liver oil and hypophosphites, and ordered to Hyères for the winter. She passed two successive winters at Hyères with general and local improvement. She was still, however, liable to persistent cough, and a return of the pyrexia, and the physical signs remained the same. She then passed a winter in England and decidedly lost ground, growing thinner, with occasional evening pyrexia, and Dr. Butler, of Cromwell Road, detected some signs of commencing disease of the right lung, which at a later date were confirmed by my observations.

October 6th, 1882.—She has remained stationary during the summer ; the cough continuing moderate and there being signs of tuberculosis at both apices. Chest measurement shows a circumference of 30 inches at the level of the third rib. Weight 9 st. 2½ lbs. Ordered to winter at St. Moritz.

May 22nd, 1883.—Returned from St. Moritz vigorous and well bronzed, having taken plenty of exercise, skating, walking, and tobogganing. She has lost all cough and gained much strength. At first she decreased in weight, but eventually gained up to her old standard. Comparison of chest measurements shows an increase of an inch in circumference. The whole thorax is hyper-resonant and no physical signs of consolidation can now be detected. Patient much troubled with piles, for which I recommended an operation which proved entirely successful.

1894.—Has remained in England since and is free from chest symptoms, though she is rather thin.

In this case great improvement took place at Hyères during two winters spent there, but the disease was not arrested, and increased the following year. During one winter's residence at St. Moritz, complete arrest of the disease was accomplished with all the characteristic changes, as shown in the widening of the thorax, the expansion of the lung, and the disappearance of all physical signs, and the fact that there has been no relapse during the ten years which have elapsed since her return from high altitudes furnishes strong evidence that the arrest is complete.

It may be interesting to compare the results of other physicians.

Dr. Herman Weber told us, in discussing a paper of mine read at the Royal Medical and Chirurgical Society, 1888, that out of 106 phthisical patients he sent to high altitudes, 38 were cured, either permanently or temporarily, 42 greatly improved, 16 were stationary or but slightly improved, and 10 deteriorated. Of these 106 cases, 70 were in the first stage, of whom 36 were cured, 28 greatly improved, 11 were stationary, and 6 deteriorated. Of the 36 second and third stagers, 2 were cured, 14 improved, 12 were stationary. and 8 deteriorated. These results, though as the general and local improvement is massed into one category it is difficult to make an accurate comparison, on the whole closely resemble my own results, but the cases, as they include a smaller number of second and third stage patients, are more favourable ; like mine, they present a very large percentage of improved for first stage cases.

Dr. Fisk,[1] of Denver, has furnished statistics of 100 phthisicals residing at Denver, only 4 of whom were females, and his results are : 67 per cent. more or less improved, 7 per cent. worse, 26 per cent. of deaths, and he calculates on 2 out of every 3 patients improving in Colorado.

Dr. Solly,[2] of Colorado Springs, extracted from his note-books 141 cases of phthisis, and did me the honour to frame his statistics exactly on the same lines as mine. His cases were, altogether, less favourable than mine. The patients had a longer history of illness, had more pyrexial complications, and contained more cavity cases and fewer first stagers. His results were general improvement in 67·3 per cent.; deterioration in 32·6 per cent.

Dr. Denison,[3] of Denver, published some years ago statistics of 202 consumptives residing in Denver, for an average of 1 year 9 months (a longer stay than that of my patients). They consisted of 148 males and

[1] President's Address to Colorado State Society, 1858.
[2] *Sanitarian*, February, 1891.
[3] *Rocky Mountain Health Resorts.*

54 females, with an average history of two years' illness before climatic treatment; 37 per cent. were in the first stage and 63 per cent. in the second and third stages; 56 per cent. had both lungs affected. These patients were far less favourable than mine, but Dr. Denison's results were 69 per cent. improved, 12 per cent. stationary, and 28 per cent. worse—a very successful result, and confirming the above.

To return to my high altitude cases, they appear to warrant the following conclusions:

1. Enlargement of the thorax takes place unless opposed by the growth of fibrosis or by extensive pleuritic adhesions.

2. Males and females seem to do equally well, and profit most between the ages of 20 and 30—males over 30 and females under 20 benefiting least.

3. The climate is specially beneficial in hæmorrhagic cases and in hereditary cases, and appears in the latter class to exercise a distinctly counteracting influence on the development of phthisis.

4. It is most effective in cases of recent date, though of utility in those of long standing, and, to insure its full benefit at least six months', and in many cases two years', stay is desirable.

5. With regard to the actual results of the climate, it undoubtedly produces great improvement in 75 per cent. of cases of phthisis generally, and in 43 per cent. it causes more or less complete arrest of the tuberculous process.

Its beneficial influence is best shown in tuberculous consolidation, where improvement may be looked for in 87 per cent. and arrest in 57 per cent.   Arrest takes place in 16 per cent. of the patients with excavation, and great improvement in 55 per cent.

Residence at high altitudes causes hypertrophy of the healthy lung tissue and local pulmonary emphysema around the tuberculous lesions, giving rise, in due time, to thoracic enlargement.   It is possible that arrest of tuberculous disease is partly owing to the pressure exercised on the tuberculous masses by the increasing bulk of surrounding lung tissue, which, by emptying the blood vessels, promotes caseation and cretification of the tubercle.

These changes are accompanied by general improvement in digestion and assimilation, the cessation of all symptoms of disease, the return of natural functions, as the menstrual in females when suspended, gain of weight, of colour, and of muscular, respiratory, and circulatory power.

RELAPSE AFTER LEAVING HIGH ALTITUDES

There is one important question often asked with regard to cases of arrest by mountain climates, and which is much discussed in the United States.   When can they return to low levels, or must they always reside at high altitudes ?   The opinion of most of the American doctors is that for most patients it is necessary to spend years before quitting the high altitudes, and that often permanent residence is the only safeguard against relapse.   Dr. Knight,[1] of Boston, insists on

[1] *Sanitarian*, March, 1891.

a year's residence in Colorado after all symptoms have
passed away, but he advises the hereditary cases
never to leave. Dr. Solly [1] thinks 50 per cent. of the
arrested phthisis cases may return. Certain it is that
very disastrous results have followed an early return to
the damp eastern states, and that a large number of old
consumptives have settled down altogether in Colorado.
and made it their home. As a rule they are an
active, energetic community, and are reported to form
a considerable proportion of the population of Denver.
which numbers 150,000. The medical faculty of that
city includes some fine specimens of cured tuberculosis.
and at a dinner given in my honour by Dr. Denison,
of 10 present the average weight was 200 lbs., 8 of
the guests being cases of arrested phthisis. At
Colorado Springs the beautiful Cascade Avenue is
lined with villas built by consumptives who have
found their health and home there. There is a curious
fact about these people. Colorado Springs has been
in existence twenty-one years as a health resort for
consumptives, who have not infrequently intermarried.
but in all the families of phthisical intermarriages—and
there have been many—as yet no tuberculous cases
have appeared among the offspring.

But to return to the question of relapse after quitting
high altitudes, judging by my cases, which were chiefly
treated at Davos and St. Moritz, relapse is rare, and
I have only been hitherto able to trace 20 cases of
relapse after leaving the mountains among the 91
arrests. I have kept nearly all the patients tabulated

[1] *Op. cit.*

in view, and, if I do not see them, can generally obtain information either from themselves, their relatives, or their medical men.

The only explanation I have to offer of my favourable results is the large preponderance among my patients of first stage cases, demonstrating the advantage of using the high altitude climate before the disease has attacked and destroyed large portions of the lungs, and the strength and vigour of the individual has been sapped by disease. I noticed two causes which seriously interfered with the steady progress usual at high altitudes, one being the use of the Koch treatment, which was vigorously pursued during the winter of 1890–91 at Davos, and proved disastrous to several patients, and the other was the influenza, which prevailed very severely at St. Moritz and Davos for two winters, and unquestionably killed a great many of the more weakly patients.

HIGH ALTITUDE STATIONS

1.—*Alpine Resorts*

The stations of the Grisons, St. Moritz (6,000 feet), Davos (5,200 feet), Wiesen (4,771 feet), are all well-known sanitaria and in every way suitable for winter and summer residence. The climate in winter is cold, with sufficient sunshine to allow of sitting out of doors in the sun's rays, but in the shade it generally freezes. At night the temperature falls far below the freezing point. The mean temperature for the winter

varies from 26° to 30° F. The number of rainy or
snowy days in winter is 50. Relative humidity is
about 40 per cent. Arosa (6,100 feet), in a well
sheltered position on the slopes of the Tschuggen,
stands high above its beautiful valley, and is only
5½ hours by diligence from Coire, and fully deserves
the commendation bestowed on it by Dr. William
Ewart; it is well provided with hotels.

Observation has shown that high altitude stations
should be built on the side of a mountain rather than
in a valley, as with the proper exposure they will
thus secure a larger amount of sunshine, and greater
freedom from vapour, which often overhangs the
bottoms of the valleys and specially those traversed
by streams or rivers.

In this respect, St. Moritz, Arosa, and Wiesen
enjoy an advantage over Davos, which is mainly
situated in the valley of that name ; but Davos on the
other hand, is more accessible, being connected by
rail. Of the two Davoses, Davos Platz is the chief
centre for hotels and pensions, but Davos Dörfli in
the winter receives the sun's rays two hours earlier,
though it loses them earlier also. Much has been
said about the overcrowding and consequently diminished salubrity of Davos, but for this there exist
no grounds whatever. The valley is a wide one, and
the hotels and houses are scattered over a considerable
tract of country. The valley is capable of holding
treble its present population without danger, and in
some of the hotels fair ventilation has been secured.
The drainage is excellent.

Maloja (6,000 feet) with its admirable Kursaal, is only kept open in summer now, and therefore not so available as formerly. Andermatt (4,738 feet) on the St. Gotthard route, and Leysin (4,757 feet), in the Canton de Vaud, are also utilised as winter resorts. Leysin, though not as high as Davos, is especially favoured in its site ; for, sheltered from northerly and north-easterly winds by the Tour d'Ai, and nestling among the pine woods on its slope, it stands facing south at the intersection of the two valleys of the Rhone and des Ormonds, and having no mountain sufficiently near to intercept the sun's rays, it enjoys more sunshine in winter than most high altitude sanitaria, which is highly advantageous. Added to this, Leysin possesses a well-built hotel, is furnished with the most approved appliances for heating and ventilation, and is very accessible.

The characteristics of the Alpine winter climate are great cold, combined with freedom from winds and fogs, consequently well borne by even delicate patients, who constantly in mid-winter sleep with their windows open and walk and sit in the sunshine daily.

Snow lies on the ground for three to five months, and the climate promotes abundant exercise, chiefly of the Canadian varieties, such as sleighing, skating, and tobogganing, which can be used under proper medical supervision. In April, or before, the snow melts, the atmosphere becomes exceedingly damp and the roads slushy, and invalids descend to a lower level for at least a month, returning in June for the

summer, during which season there is more wind and
rain, but not much dust. The intermediate halting
places most in vogue are Seewis (2,985 feet), Thusis
(2,448 feet), Ragaz (1,709 feet), Gais in Appenzell
(2,820 feet), where patients remain until they leave for
home or return to the mountains. The good results of
these Alpine climates may be judged of by the above
statistics of patients, the greater number of whom
resided in the Grisons.

2.—*South Africa*

The South African mountain stations, where a
considerable number of my patients have lived with
benefit, are situate in the Cape Colony, the Orange
Free State, and the Transvaal, at an altitude of from
4,000 to 5,000 feet, the conformation of the country
being that of rolling veldt. Taking Bloemfontein, the
capital of the Orange Free State, as a type, the
climate is warm, with a few cold nights in winter.
The average minimum is 55° F., the average maximum
for the six hot months 82° F., the relative humidity 55
per cent., and the rainfall 17 inches. The number of
rainy days 70. It is a splendid climate, and has
yielded excellent results in my hands and in that of
others, and Cape railways and steamers have rendered
the whole district fairly accessible, though the accommodation for invalids is not up to the European standard.
In Cape Colony, Aliwal North, Beaufort West, and
Matjesfontein in the Great Karoo district, Tarkastad,

Dordrecht, and Burghersdorp, and Lemoenfontein on the Nieuwveld mountains, have been established as health resorts, and there will be an extension in this direction when required.

The Cape Colony and the adjoining Orange Free State and Transvaal contain vast tracts of country at a considerable altitude above sea level, which from their dry sunny climate have been found most beneficial in the treatment of lung diseases. The conformation presents a comparatively low belt of country fringing the coast and rising into plateaux and mountains of considerable elevation on proceeding inland. Here we have the Zwartebergen, Nieuwveld, the Drakensberg and other ranges, one of these, the Compassberg (7,800 feet) being the highest mountain in the Cape Colony. A large portion of this elevated ground consists of veldt, or rolling prairie, varying from 2.000 to 5,000 feet in height.

The healthiest regions of the Cape Colony are the Central and Upper Karoo districts, the former lying between the Roggeveld and Nieuwveld mountains on the north, and the Zwarteberg range on the south, and extending west and east from the Hantam in Calvinia district to the Sunday River in the Graaff Reinet district, *i.e.* over 5 degrees of longitude, with an average altitude of 3,000 ft. The latter, or Upper Karoo plateau, includes the tract of country lying north of the Nieuwveld mountains and south of the Orange River, and is bounded on the east by the Orange Free State and on the west by Namaqualand, embracing the districts of Aliwal North, Albert, Colesberg, Victoria West,

Kimberley and others, and varying in height from
2,700 to 6,000 ft.

The climate of the Great or Central Karoo,
according to Dr. Saunders, is characterised by extreme
dryness and prolonged droughts, occurring at intervals.
The summer heat is intense, reaching 110° F. in the
shade, but, like in Arizona and California, bearable on
account of the dryness of the atmosphere, while the
nights are cool.   The winds in summer are mostly
from the north and north-west. bringing heat and great
clouds of red dust. probably from the northern desert.
Thunderstorms are of great violence and follow in the
wake of the north-west winds, converting in a few
hours large tracts of country into temporary lakes: but
these are rare.   With regard to rainfall, it is undoubtedly very small. from 9 to 18·36 inches, and the
number of rainy days is inconsiderable.   According to
Dr. W. H. Ross, the gradual denudation of the soil by
bush fires, and the careless cutting down of trees. has
intensified the effect of the sun's rays and the desert
winds on the soil : the greater part of the Cape Colony
land is glazed with baked clay. from which the water
runs off as fast as it falls, and there is nothing to retain
moisture and allow of slow filtration. and except in the
neighbourhood of Knysna, the George forests and
the few miles of moderately well-wooded territory.
there is no certainty of a water supply.

The winter is characterised by cold nights and by
several hours of bright sunshine between 9 A.M. and
3 P.M.   Snow appears in the high mountains, but not
in the veldt, where fires are rarely needed, except in

the evenings ; the air is clear, bright and bracing, admitting of abundant exercise.

The railway to Kimberley and Bechuanaland runs through the Karoo, rendering the district very accessible, and consequently many towns are springing up along its line. The most suitable stations in the Cape Colony for pulmonary invalids seem to be Beaufort West (2,792 feet), Lemoenfontein (3,192 feet), Aliwal North (4,318 feet), Burghersdorp (4,552 feet), Colesberg (4,407 feet), Tarkastad (4,280 feet), and Kimberley (4,012 feet), all of which have reported cases of arrest of phthisis. Both the Orange Free State and the Transvaal lie entirely at a high level and offer most promising sites for sanitaria. Bloemfontein (4,500 ft.), the capital of the Orange Free State, is an old established health station, and now connected by rail with Port Elizabeth and Cape Town. Its climate, like that of the Orange Free State generally, differs from the climate of the Kimberley district, and the town has the advantage of being better sheltered by neighbouring hills. I have more than one case of phthisis which owes its arrest to long residence here, the only drawback being the drainage, which has been much complained of lately, but improvements in this direction are contemplated. In the Transvaal, which will soon be in railway communication with the coast and the Orange Free State, are Pretoria (4,000 feet), the capital, a well-sheltered attractive little town ; Heidelberg, also well spoken of, and Johannesburg (5,000 feet), of gold-mine notoriety, with a fine climate but a dusty atmosphere, which is now settling down into a

well-organised city of upwards of 20,000 population.
One of the many advantages offered by the South
African Highlands, the claims of which have been
strongly urged by my colleague Dr. Symes Thompson,
is that here, as in other rising and not yet over-
crowded countries, there is the possibility of a career
of usefulness being open to the patient, in the climate
where his disease has been arrested.

Case XII.—A young gentleman, *æt.* 20, belong-
ing to a very phthisical family, was seen January
16, 1874, with a severe cough and wasting, and a history of an inflammatory attack of the left lung four
years previously, for which he went to the Cape of
Good Hope and recovered, but at the diamond fields he
contracted fever and diarrhœa and was much weakened.
He returned to England and went up to Cambridge
University, where two months later he caught a severe
cold boating, and violent cough with expectoration
followed. Dulness and increased vocal resonance were
detected in the upper right chest. Cod-liver oil with
dilute phosphoric acid and quinine was prescribed,
and a return to the Cape recommended.

May 8, 1876.—Has spent two years in South Africa,
principally at Kimberley, and has apparently quite
recovered. He lost the cough on the voyage, and
proceeded at once to the diamond fields, where he
soon regained his flesh and strength, and conducted
very successful and lucrative minings. On examination
I found his chest greatly expanded, and hyper-resonant,
the heart being displaced downwards into the epigas-
trium. No signs of consolidation could be detected.

The after career of this patient was remarkable. He returned to Cambridge, took his degree, and was called to the English bar. He set out again for the Cape, where he was elected to the House of Representatives and became chairman of one of the largest and most successful diamond mining companies in the world, and now holds posts of the highest distinction and responsibility; and all this without any serious breakdown in his health.

The case may be considered a good example of (1) the beneficial effect of the Cape high altitude climate, and (2) of the lasting effects of such benefit.

3.—Colorado and the Rocky Mountains.

The Rocky Mountain climates,[1] where six of the patients resided for periods of three to eleven years, with such great benefit, present many advantages, especially to Americans. Owing to the conformation of the country, the whole of the States of Colorado and New Mexico are of altitudes varying from 5,000 feet to 11,000 feet, and even rise to 14,000 feet in the Rocky Mountains themselves.

Considering the elevation of the States of Colorado and New Mexico the climate is remarkably mild, snow seldom lying long in winter, while in summer the altitude keeps the air comparatively cool. The natural parks are well supplied with water, and some, like the Middle Park, have sulphur springs; while railways

[1] These are more fully described in the *Address on the High Altitudes of Colorado*, p. 146.

encircle, and in some cases penetrate into, them. Hotels have been established, and accommodation will shortly be improved and extended.

The winters are bright and clear, with frosty nights, but scarcely any snow, and on the prairie there is a good deal of wind. The climate on the whole may be described as very dry, and clear and sunny, very windy and abounding in electricity. This estimate is of course modified, according to the elevation, and to the position in regard to the mountain chain.

Compared with Alpine sanitaria, Colorado offers the advantage of no snow-melting season, and of allowing continuous residence all the year round in a somewhat warmer and drier climate, but it has the drawbacks of wind and dust, most troublesome in summer. With regard to results, it has not yet produced as favourable statistics as the Swiss. It presents also a great variety of altitude, so that invalids pass their summers camping out in the parks, and leading an open-air life, hunting, shooting, and fishing, while they spend their winters in Denver and Colorado Springs, or in the foothill towns. But perhaps a greater attraction still to many is that, for a large proportion of patients, Colorado offers a prospect of profitable occupation. whether professional or commercial, and the Western people kindly welcome strangers into their rising State, well knowing that there is room for all within its large confines.

The subjoined is a good example of the arresting power of this climate.

Case XIII.—A young American lady, aged 23,

sprung from a healthy stock, was seen by the late Sir Andrew Clark and myself in consultation with Dr. Brisbane on May 29th, 1878. She had suffered from cough and expectoration, night sweats and evening pyrexia, and loss of flesh for about two years, and we detected consolidation of the right upper lobe, and cavernous sounds were heard in the first interspace. This lady married an Englishman, and passed two summers in the United (Eastern) States and two winters in England, one in Brighton and one in London, but lost ground, and when Sir Andrew Clark and I saw her again in February, 1881, though we found signs of contraction of the cavity at the right apex, crepitation was heard as low as the fourth rib in front and over the whole surface of the lung posteriorly. In spite of treatment the disease extended, and in May of the same year we discovered that the left apex was attacked. After a careful consideration of the case and a survey of the attendant circumstances, we recommended a year's sojourn at Colorado Springs, for which place she started in June, 1881. Finding it too hot she ascended to Waggon Wheel Gap (9,000 feet), 30 miles further in the Rockies, and resided there for three months with decided improvement. She then returned to Colorado Springs and lived an outdoor life, renouncing drugs, eating heartily, and drinking three quarts of milk a day. She increased in weight 36 lbs., and at one time lost her cough, but later it returned, and when I saw her in 1882 she complained of laryngeal irritation. She had become well bronzed, and complained of my not recognising her, which was

perfectly true. The physical signs showed enlargement
of the whole thorax, hyper-resonance everywhere; dry
cavernous sounds were audible from the clavicle to
the third rib on the right side. With the exception
of a few fine crepitations in the right axilla, the
crepitation had disappeared, but harsh breathing was
audible over both lungs. Her walking power was
very good. I saw no more of this patient, but when
at Colorado Springs last October I heard from Dr.
Reed the sequel of the case. The disease remained
arrested for seven or eight years, the patient passing
her winters at Colorado Springs and her summers in
the ·Eastern States, but early in 1891 she went to
New York to try a bacillicide treatment, of which
starving was a leading feature, and returned to Colorado
Springs in a miserable condition with progressive
disease. She did not improve under Dr. Reed's treatment, and so was transferred to the care of a Christian
Science lady, and gradually sank in a few weeks, dying
in the spring of 1892.

The case is a sad one, but it demonstrates the
wonderful arresting power of the climate over the
onward march of phthisis, for the progress of the
disease in 1881 was such that Sir Andrew Clark and
I hesitated before recommending Colorado.

4.—*The Andes*

This magnificent range may be said to have been
the cradle of the mountain cure of phthisis, for it was
from the experience gained by Archibald Smith and

Guilbert in Bolivia and Peru, that the attention of English and French medical men was first directed to the subject, which resulted in sending consumptives to the high altitude sanitaria in Europe and in North America.

The Andes, owing to their running nearly due north and south, and extending from 10° north latitude to 45° south latitude offer a great diversity of climate, which is characterised by two principal features: (1) The remarkable dryness of the whole of their western slopes and valleys. This is owing to the very unequal distribution of the rainfall and water drainage, which is nearly wholly carried down the eastern watershed, and flowing eastwards, combines to form the great rivers of the Orinoco and Amazon and the La Plata, by which South America on that side of the Andes is watered and fertilised. On the western side, the range is more or less precipitous, the Andes, as in Peru, rising directly from the Pacific Ocean and here are few streams and little or no rainfall. In Peru itself no rain falls (Guilbert), and the country is watered by a mist, and in the northern part there is a considerable tract of desert country, producing nothing but nitrates of soda and potash, except where water is brought for irrigation purposes, and where, as might be expected, there is abundant fertility. (2) Owing to the conformation of the region a variety of climates are to be found in a small compass, and the fruits of the tropical and temperate regions are to be seen in the same market. Extensive plains exist at high elevations, on which populous cities are built, and in some of these the climate is temperate and genial. Santa Fè de Bogota, in New Granada, with a population of 40,000, stands at an elevation of 8,648 feet, with a climate like that of Malaga, and an annual mean temperature of 59° F., the mean of each season hardly varying from that figure, and including scarcely any extremes. Quito, the capital of Ecuador, with a population of 80,000, is situated on the east side of Picinchincha at 9,500 feet elevation, has a climate a little warmer than that of Bogota, and which has been compared to perpetual spring, the mean annual and seasonal temperature being about 60° F.

In Peru there is Arequipa at 9,000 feet, easily accessible from the Pacific ; and in the well-sheltered valley of Jauja lie Tarma and Jauja, at about 10,000 feet elevation, health resorts of considerable repute, the former being used by the Peruvian Government as a sanitarium for military consumptives. Huancayo, to the north of Jauja, has a climate intermediate between Tarma and Jauja.

In Bolivia we have the capital of La Paz (population, 78,000), at a height of 13,500 feet, with a more bracing climate than the above mentioned. All these towns have tolerable, and in some cases excellent, hotels, and are connected with the Pacific ports by roads, and in some cases even by rail. The famous Oroya line, which crosses the Andes at a height of nearly 17,000 feet, nearly reaches Tarma and Jauja and connects them with Callao, the port of Lima. Communication with England is by the Royal West India mail steamer to Colon, thence by rail to Panama,

and from this point by steamer to Callao, the journey
from London lasting about six weeks. The advantages
of the Andean climate are the combination of warmth
and equability, with rarefaction, and the striking effects
of its influence on consumptives are to be seen in
South America, also a few patients sent from Europe
have testified to its beneficial results.

The drawbacks are, the distance from England,
the long and fatiguing journey, the possible ascent of
passes of great altitude (some of these of 16,000 feet
and upwards) and consequent suffering from mountain
sickness, and the Spanish food and cooking; so that
the Andes can only be considered a fit resort for
energetic young men, with limited tubercular lesions,
capable of enduring fatigue, and able to accommodate
themselves to conditions of life unlike those to which
they are accustomed.[1]

5.—*The Himalayas and other Indian Ranges*

The Himalayas afford a variety of altitudes in the well-
known hill stations, which vary in height from 4,000 feet
to 8,000 feet and are largely utilised by the Indian Government as sanitaria for European troops. The climate is
a remarkably fine one, the mean temperature ranging
from 40° F. in winter to 78° F. in summer. The rainy
season lasts from June to September and the rainfall
is heavy; from 70 inches at Simla to 132 at Darjeeling.

---

[1] For further particulars, see *Influence of Climate on Consumption*,
by Author, p. 112.

The most reputed stations are Landour (7,300 feet),
Murree (6,786 feet), Kasauli (6,400 feet), Naini Tal
(6,200 feet), Simla (8,000 feet), and Darjeeling (8,000
feet).   The Himalayan climate, except during the
rainy season, is a very invigorating one, but somewhat
hotter than that of England.   The testimony of Indian
medical officers appears to be rather against the use of
the Himalayas in the treatment of lung diseases,
though they are beneficial in other maladies, and
apparently this is on account of the heavy rainfall,
which, during the summer, causes a very damp atmosphere, the opposite to what we have in the Andes and
Rockies.   Another objection is the conformation of the
country, which consists of more or less precipitous
rocks and deep valleys separating them, the mountain
plateaux being small and not extensive enough to
admit of sufficient exercise being taken on the level.

The Nilgiri mountains in the Presidency of Madras
range from 5,000 to 7,000 feet in height and enjoy a
more equable and dryer climate than the Himalayas.
These are much utilised by Europeans in India.   The
winter mean temperature is about 60° F. and the
summer mean about 65° F., the rainfall 55 inches.
with occasional fogs.   The principal stations are Ootacamund (7,361 feet), Kotagherry (6,100). Wellington
(5,840 feet), and Coonoor (5,161 feet).

The Pulneys Hill station (7,000 feet). the most
southerly hill station in India, was strongly advocated
by the late Dr. John Macpherson, partly because its
position on the Indian peninsula secured for it the
influence of salt breezes from both the Cochin and
Tutacorin coasts, and partly the rainfall is small.

EFFECTS OF HIGH ALTITUDES IN OTHER DISEASES

Before quitting high altitude climates I would mention in what other diseases besides phthisis they prove beneficial :

1. Imperfect thoracic and pulmonary development.
2. Chronic pneumonia without bronchiectasis.
3. Chronic pleurisy where the lung has not expanded after removal of the effusion.
4. Bronchial asthma without emphysema.

These climates are contra-indicated in :

1. Phthisis with double cavities.
2. Fibroid phthisis, and all cases where the pulmonary area at sea level hardly suffices for respiratory purposes.
3. Catarrhal and laryngeal phthisis.
4. Acute phthisis of all kinds, and especially when there is great irritability of the nervous system.
5. Phthisis with pyrexia, and in pyrexial cases generally.
6. Emphysema.
7. Chronic bronchitis and bronchiectasis.
8. Diseases of the heart and great vessels, except functional ones : diseases of the liver and the kidneys, including all forms of albuminuria (Andrew Clark).
9. Diseases of the brain and spinal cord, and conditions of hypersensibility of the nervous system.
10. Anæmia.
11. In patients of advanced age, or too feeble to take exercise.
12. In patients whose organs have become degenerated by long residence in tropical climates.

HOME STATIONS

We have discussed the element of barometric pressure at some length and time will, I fear, not suffice to enter upon the remaining elements enumerated—namely, *wind-force* and *atmospheric electricity*, but they are reserved for a future communication. I propose now to make some brief comparisons between the climate results already presented and those of our own home stations. I had not intended introducing these last at all, because they were treated at some length in my Lettsomian Lectures published in 1877,[1] but I have, for the purpose of comparison, extracted the principal results of these 243 consumptive patients, adding 49 fresh cases, making in all 292, and have classified them with the same headings as the others in Table IV :

TABLE IV.—*Results of Different Climates Compared.*

	No. of Patients.	Average Length of Residence.	First Stage.	Second and Third Stages.	Bilateral Affection.	Results.						
						General.			Local.			
						Improved.	Stationary.	Worse.	Arrest.	Improved (including Arrest).	Stationary.	Worse.
		Months.	Per Cent.	Per Cent.	Per Cent.	Per Cent.	Per Cent.	Per Cent.	Per Cent.	Per Cent.	Per Cent.	Per Cent.
High altitudes ...	247	12·2	65·0	35·0	37·0	83·4	2·02	14·57	42·5	75·5	5·3	19·1
Sea voyages	65	1·6 (average of voyage).	63·0	37·0	37·0	77·0	—	21·56	7·7	53·3	10·7	33·8
Riviera	210	9·0	59·0	41·0	36·0	65·2	10·00	24·80	5·9	36·6	17·8	45·6
Home climates...	292	9·7	58·0	42·0	42·0	63·7	8·21	28·00	2·0	38·9	20·0	41·1

This shows that the largest number of patients tabulated, tried the home climates, and the smallest, sea

[1] *Influence of Climate in Consumption.*

voyages; and that the average period of residence was, for the Riviera, nine months; for the home climates, nine months and a-half; and for the high altitudes, twelve months; though as I showed when dealing with the statistics, ten months is nearer the real average of these last, the difference being made up of cured patients who have become residents. The sea voyages are almost impossible to reduce to an average of months, so I have had to make a rough one of voyages, the average being 1·6. The next three columns deal with the lung condition, and give the relative percentages of patients in each category, which do not differ widely, and between the highest and lowest percentages in the first and second and third stage categories there is only a difference of 7 per cent.; in bilateral affections only 6 per cent. The high altitude cases seem to be rather the most favourable, and the home climates rather the least, but not markedly so.

The results are very striking. In *general* results the home climates yield the smallest percentage of improvement, and the largest of worse; next comes the Riviera, not much better; then, with a rise of 12 per cent. improved, are sea voyages, the percentage of worse being still large. High altitudes win easily in all categories, with their 83 per cent. improved, and only 14½ per cent. of worse.

In *local* results the arrest cases have been made a separate division, but they are also included under the improvement percentage. We see here that the Riviera comes out worst, except for a large number of arrests. Next we find home climates, then separated

by an interval of 14 per cent. more of improved, and
12 per cent. less of worse, are sea voyages. The high
altitudes again come out *facile princeps* in all categories
with favourable percentages, nearly double those of the
Riviera and home climates. Looking calmly at these
results, it must be admitted that there is strong evidence
in favour of high altitude treatment.

The table does not profess to be perfect, but, considering the number of cases included and the average
length of residence, it affords some fair grounds of
comparison.

Our comparisons would be still more instructive if
we could be certain that the patients pursuing different forms of climatic treatment conformed to the
same rules of hygiene and dietetics, which is not
always the case. Undoubtedly careful medical supervision should be carried out in all these cases, and one
reason of the success of the high altitude treatment, as
practised in the Alps, is that such supervision is easier
and more complete than in the Riviera and in southern
resorts. The success which has attended Brehmer at
Görbersdorf and Dettweiler at Falkenstein in their
Sanitaria, has been due to the great care bestowed on
the supervision of the patient's life and habits, rather
than to any special merit of the climate of these
localities, and this should be borne in mind in sending
patients to the various health resorts.

THE HIGH ALTITUDES OF COLORADO AND THEIR CLIMATES.[1]

Mountain Climates —Varieties — Distribution of Altitude in the United States — Rocky Mountains — Topography — North, Middle, South, and San Luis Parks—Origin of Parks—Cañons —Clear Creek—Arkansas—Great Cañon of the Colorado—The Chains of the Rockies in Colorado—Gray's, Long's, and Pike's Peaks—Contrast to Alps—Geology and Mineralogy—Four Series of Elevations—Snow-clad Peaks—Meteorology of Colorado—Rainfall—Relative Humidity—Cloud—Rainy Days chiefly in Summer—Mean, Maximum, and Minimum Temperatures—Great Range—Wind Prevalence and Force—Large Number of Fine Days—Cause of Dryness of Climate—Hours of Sunshine—Pike's Peak—Meteorology—Relation of Atmospheric Pressure to Air Temperature—Wind Velocity—Electrical Phenomena—Climate of the Natural Parks— Estes Park—Elevation and Remarkable Position—Shelter—Dry and Sunny Climate—Manitou Park—Climate of the Foothills— The Prairie — Atmospheric Effects — Vegetation — Denver— Longmont and Hygieia—Boulder—Greely—Fort Collins— Idaho Springs—Colorado Springs—Meteorology—The Invalid's Day—Charming Excursions—Manitou Springs—Elements of the Climate of Colorado.

MOUNTAIN climates, though classed together, vary in temperature, moisture, and wind prevalence, according to latitude, distribution of land and water, and relation

[1] A Presidential Address delivered to the Royal Meteorological Society, January 18th, 1893.

to other mountain ranges ; but they have one common feature—to wit—diminished barometrical pressure, varying with the altitude, and itself giving rise to another peculiarity of mountain climates, diathermancy, which greatly influences animal and vegetable life ; of this more will be said presently.

Naturally, too, we should expect that the atmosphere of mountains should be cooler than that of plains, and being cooler, it is incapable of holding in suspension as much moisture, and is therefore drier. Here again the position of the range comes in, for if, as in Assam, the mountain lies in the track of a wind laden with moisture, such as the south-west monsoon, the mountain causing the current to ascend, serves as a condenser, by lowering the temperature, and we get an annual rainfall, like that of Cherapunji, of 493 inches. If, on the other hand, it lies like the American Rockies to the lee of other ranges, such as the Sierra Nevada and Wahsatch mountains, the rainfall is moderate, and the climate dry.

There is another class of phenomena which is very remarkable in mountain climates, viz. the electrical, for from their position, mountains are the natural conductors of terrestrial and atmospheric electricity, and it is no wonder that they are the scene of marvellous displays, as we shall hear when we allude to Pike's Peak.

The climate of Colorado is my text to-day, first, because I have lately returned from an interesting visit there, undertaken for the purpose of investigating the leading characteristics of the Rocky Mountains

and secondly because the situation of this chain in the
heart of a great continent, far away from the sea or any
large body of water, renders it especially fitted to
serve as a type of a mountain climate. The annexed
chart (Diagram 9), founded on a relief map of the
United States, supplied to me free of cost by the Geo-
logical Survey at Washington, with that liberality which
distinguishes the United States Government in all its
scientific relations, will show that west of the Mississippi
valley a gradual rise takes place, which is still more
marked when we reach longitude 99° west, the line of
elevation extending from north to south of the United
States, and here a height of from 2,000 to 5,000 feet
is reached. This comprises the vast prairie lands of
North and South Dakota, Nebraska, Kansas, and part
of Texas, a great portion of which are now cultivated
farms. West of this we observe a further elevation,
limited chiefly to the States of Wyoming, Colorado,
and New Mexico, of from 5,000 to 8,000, 11,000, and
even over 14,000 feet in the Rockies themselves. It is
in these three states that we find the localities which
are utilised, or likely to be utilised, as health resorts.
The Rocky Mountains run nearly north and south,
through British North America and the United States,
and consist of a more or less distinct central chain and
spurs running in various directions.

In the United States the range appears to bear
different names in different states. In Montana it is
called the Rocky Mountains, in North Wyoming the
Bighorn Mountains, in South Wyoming the Black
Hills ; the spur running transversely from the main

RELIEF MAP OF THE UNITED STATES.

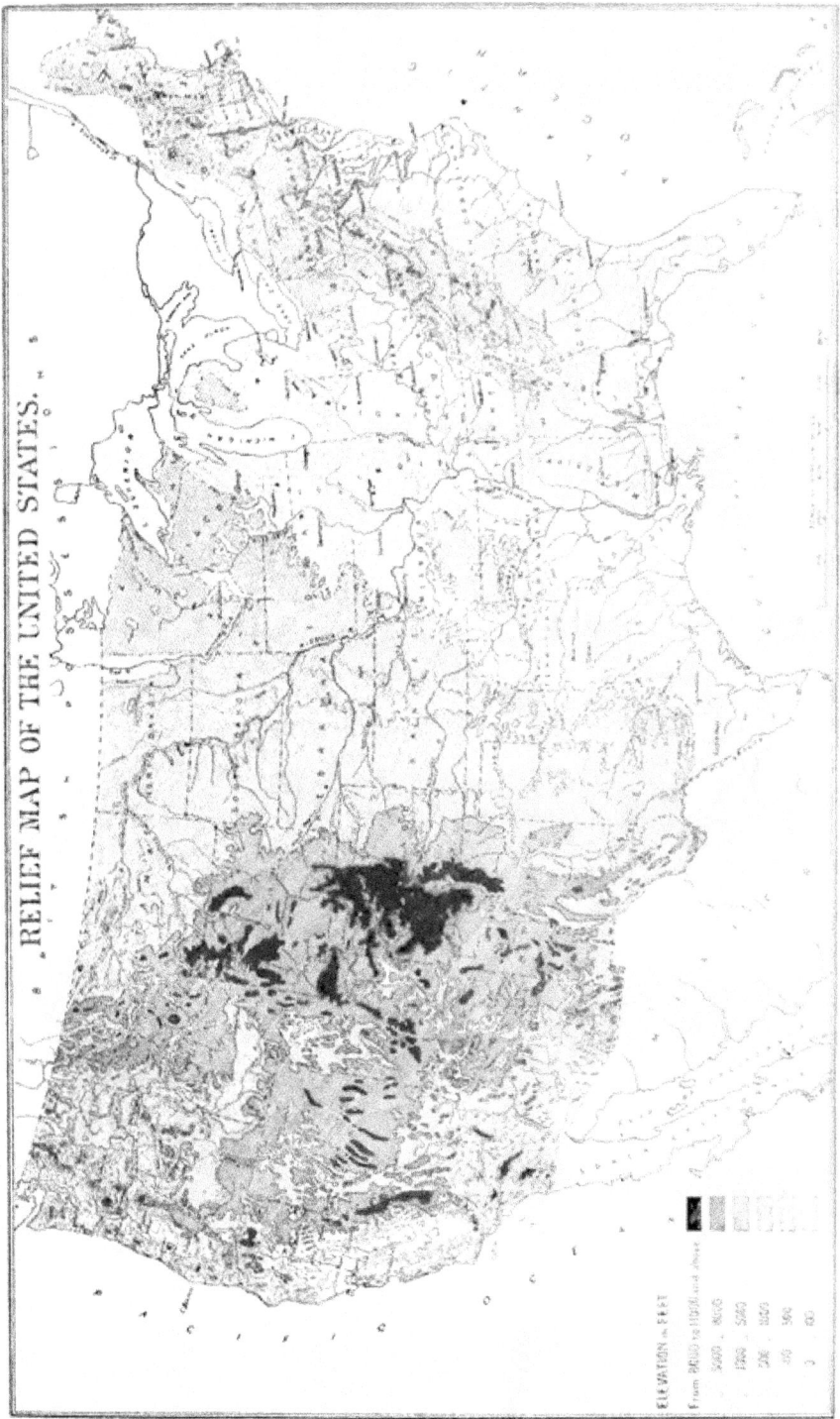

ELEVATION in FEET

From 8000 to 10000 and above
5000 – 8000
2000 – 5000
500 – 2000
0 – 500

Malby & Sons lith

chain into the territory of Utah is called the Wahsatch Mountains, a portion of which is known as the Uintah Mountains and forms, with the last mentioned, a feature in the beautiful view from Salt Lake City. The range extends over a large portion of the State of Colorado, is penetrated by various valleys and streams, including the famous Rio Grande, and here rises to its greatest elevations; most of the grand peaks, Sierra Blanca, Mount Lincoln, Long's and Pike's and Gray's Peaks being situated within the confines of this State. The main range ends in the beautiful Spanish Peaks, but an offshoot, passing to the west, extends into Arizona, a large proportion of which state is elevated ground.

In the state of Colorado are various chains of the Rockies, two of which, running north and south, are called Front and Park ranges. These enclose between them the great natural parks—North, Middle, South and San Luis Parks—which are wide valleys of pasture land at a high altitude, sheltered from winds, and surrounded by beautiful scenery. North Park stands at an elevation of 8,000 feet, and contains 2,500 square miles. Middle Park, which has a warmer climate, is 65 miles from north to south, by 45 miles from east to west, and embraces about 3,000 square miles, 7,500 feet above sea level. Communicating with it are Estes Park, Antelope and Manitou Parks, to which we shall refer presently.

Middle Park is enclosed by grand mountains, Long's and Gray's Peaks and Mount Lincoln; it contains beautiful meadow land and some celebrated sulphur springs. South Park is 60 miles long and 30 miles wide, including 2,200 square miles of, for the most part, pasture land, at an elevation of 9,000 feet, and it has a milder climate than the other two great parks. South of these is the San Luis Park, larger than all of them together, and containing 18,000 square miles with an elevation not exceeding 7,000 feet, and its more southerly latitude, combined with a very abundant water supply, causes it to be more thickly wooded than the others.

All these parks enjoy fine climates ; the wilder parts abound in elk, deer, mountain sheep, antelopes, bears, wolves, lynx, mountain lions, coyotes, and beavers, and the streams supply plenty of trout. Cattle pasture all the year round without shelter or covering, and thrive on the abundant grass. It is supposed by Dr. Denison [1] that these parks were once beds of immense bodies of water, lakes in fact, which breaking through their rocky barriers cut deep rugged gorges or cañons, through which the rivers have flowed for centuries, depositing their *débris* in the foot-hills and plains.

The deep gorges, or cañons, are a remarkable feature of the Rockies, as they may be found penetrating the chain in many directions. They are clearly the result of water action, and the various layers of strata cut through and exposed are well shown in the neighbourhood of Denver, where we have a good example in Clear Creek Cañon, where the river, starting from the base of Gray's Peak, has forced its way through a labyrinthine channel eastwards to the plain at Golden.

[1] *Rocky Mountain Health Resorts.*

A grander cañon is that of the Arkansas, which is the line taken by the Denver and Rio Grande Railway in penetrating the Rockies. In parts, owing to the narrowness of the gorge and the great height and precipitous character of the giant granite cliffs and mountains through which the Arkansas threads its way, the scenery is magnificent. The forms of the rocks are specially fine, a specimen of which is seen for instance, in the Currecanti needle, a noble pyramid of granite, and the mountain of the Holy Cross, which owes its name to the form of the Cross being delineated by snow accumulating in crevices in its rocky sides. The grandest piece of scenery on this route is the Royal Gorge, where the chasm is so narrow that there is only space for the river to flow through, and the railroad is therefore suspended by an ingenious bridge above it for some distance, the cliffs on either side reaching the height of 2,600 feet. Another striking mass of rocks is the Castle Gate, the fine portal by which the railway escapes from the gorges into the more open valleys on the western side of the Rockies.

This cañon of the Arkansas is surpassed in grandeur and wild magnificence by the great cañon of the Colorado situate in Arizona, where that river boils and surges at the bottom of the greatest cañon in the world, one mile in depth, with nearly precipitous and beetling cliffs on either side ; but this is outside of the state of Colorado.

The railways encircle, and in many cases penetrate into, the region of the parks. The Denver and Rio

Grande sends branches up several of the smaller cañons opening into them, while Middle Park is traversed by both the Denver and Rio Grande and by the Midland of Colorado, so that the district is rendered fairly accessible.   Moreover the numerous towns, for the most part mining centres, like Leadville, scattered throughout the Rockies or situated like Denver on the adjacent plains, amply supply the wants and necessities of residents and of camping-out visitors.

The peaks of the Rocky Mountains vary in altitude from 13,500 feet to 14,500 feet, the highest being the Sierra Blanca, which terminates the Sangre de Cristo chains; there being more than 100 peaks exceeding 13,500 feet in height, and Gray's and Pike's and Long's Peak all exceed 14,000 feet, the general average being higher than in the Swiss Alps.

In appearance, however they are not nearly so imposing as the Alps; they rise from a higher plateau varying from 5,000 to 6,000 feet in height, their outlines are more rounded, the lower slopes, probably on account of the great dryness of the atmosphere, are less wooded, and for the same reason, and also because of the lower latitude, the summits have less snow massed on them and no glaciers furrow their sides. The absence of cloud and mist causes their outlines to be very clearly defined even at great distances, and they form a fine feature at the ends of the tree-lined avenues of both Denver and Colorado Springs.

The geology and mineralogy of the Rocky Mountains are too complicated subjects for me to deal with, (especially as, according to the Denver and Rio Grande prospectuses, all known minerals are found there!) but the greater part consists of masses of granite rising beyond an outer stratum of red sandstone, which last forms the very weird and picturesque rock scenery of Monument Park and the Garden of the Gods, among other instances. The colouring of both granite and sandstone is very vivid and rich in tone.

The Rockies, as will be seen by our maps, form the greater part of the State of Colorado, the eastern portion consisting of the prairie, which extends far to the east and north, and in this State has an altitude of about 5,000 feet. Between Denver and Colorado Springs a spur of the range runs eastwards into the prairie, called the Divide, in which lies Castle Rock, Perry Park and Palmer Lake, but south of this the great plain extends to the Gulf of Mexico, 800 miles.

We must therefore remember that in Colorado State there are four series of elevations, viz. :—

1. The snow-clad peaks, 12,000 to 14,000 feet and upwards in height, with a climate of their own.

2. The natural parks, varying from 7,000 to 10,000 feet in altitude.

3. The foot-hills and adjoining valleys, rising from 6,000, to 7,000 feet.

4. The prairie plains, varying from 5,000 to 6,000 feet.

The first class is of course useless for health station purposes, but the last three might be made available and offer an infinite variety of altitude and of climate.

Having described the general conformation of the

country, we come to its meteorology as a whole, and the annexed map of the Weather Bureau of the United States will give some idea of the distribution of the rainfall.

DIAGRAM 10.—MEAN ANNUAL RAINFALL OF THE WESTERN STATES.

In Colorado the rainfall varies from 8.71 inches to 22.30 inches, at Denver it is 14.17 inches, at Colorado Springs 15.17 inches, and at Santa Fé, the capital of

New Mexico, 14·17 inches. It does not appear to increase with altitude, for Gunnison, 7,680 feet, in the heart of the mountains, has only 10·02 inches, and Leadville, at 10,200 feet, only 12·80 inches. Georgetown has 14·39 inches ; Pike's Peak, with its great elevation and consequent liability to act as a condenser, has a larger rainfall, viz. 29·18 inches, but this is after all small in comparison with other high peaks. The western stations, such as Fort Lewis, seem to have larger rainfall than the eastern.

Throughout Colorado scarcely any rain falls from October to April, and the greater part is precipitated during the thunderstorms, which are so frequent from May to September. Snow occasionally falls in autumn and winter, but except at the higher levels does not lie.

Relative humidity returns from several portions of the state show a percentage varying from 46 to 58 per cent., contrasting with those of the Pacific coast, such as at San Francisco 78, and at St. Louis, in the valley of the Mississippi, 78, or on the Atlantic, as exemplified by New York, 75 per cent.

The average amount of cloudiness in Colorado, as given in Dr. Denison's charts, is also very small ; whether the season be spring, summer, autumn or winter, it always shows a percentage far less than that of either coast line. The number of days on which rain falls is about 85, but on many of these the day is fine generally, with an evening storm.

The temperature shows great extremes, and, as might be expected in a mountainous region removed from all equalising influences, such as that of the sea, the nocturnal radiation is considerable. The mean annual temperature of Denver is 50°; the maximum reading, occurring in July, being 100°, and the minimum, 7°, in February.

The monthly means are as follows :—

January	...	27·2	July	72·0
February	...	29·6	August	72·8
March	...	43·3	September ...	60·0
April	...	51·1	October	51·8
May	...	55·5	November ...	32·4
June	...	64·3	December ...	40·5

The summers are warm, tempered by showers, as most of the rain falls in summer and spring. The autumn is fine with, perhaps, some frosty nights, but very sunny and warm by day. There is practically no winter till January, and then but little snow, which the powerful sun usually soon melts, though nocturnal frosts are frequent and very severe, but the striking feature is the range of temperature, which amounts at Denver occasionally to 107°, and has been known in some stations of Colorado to reach 118°.

The minimum has been known to fall to −25°; this was in January, 1876, and 101° is the highest maximum recorded. Wind is undoubtedly present and may be almost said to be a feature of the climate, though some spots are completely protected. The wind prevailing is the south, and the average wind force for the year is 7. Only between 30 and 40 days are reported as absolutely calm, so the climate may be considered as by no means devoid of aerial movement.

The number of fine days is one of its strongest recommendations, and I see the Weather Bureau for 1889 records for Colorado Springs 171 cloudless days, 135 partly clouded ones, and 59 cloudy.

These figures all point to a very dry climate, with small rainfall and relative humidity and great absence of cloud and vapour ; and a glance at the relief and rainfall maps will explain this peculiarity. The usual rain-bringing winds come from the Pacific coast, where it will be noted that the rainfall is comparatively large. but before they reach Colorado State, and more especially the parks and plains, they are driven upwards over the Sierras, which condense much of their moisture ; next they have to cross 60 miles of the great American desert, which is not likely to add to their humidity ; then they are deviated upwards again by the Wahsatch Mountains, and passing at an altitude of at least 13,000 and 14,000 feet over the Rockies, they arrive at Colorado as dry, and for the most part warm, winds. The country to the east is a vast plain gradually descending to the Mississippi valley, and on the whole dry, the only source of moisture being the Gulf of Mexico, 800 miles off ; the consequence is that Colorado and New Mexico enjoy remarkably dry and cloudless climates, and are also extremely sunny regions. The hours of sunshine far exceed in number those which can be counted at most European resorts. Dr. Solly gives a table of temperatures indicated by the solar radiation thermometer during the year 1886—7, which conveys a fair idea to the mind of the reader of the warmth and brightness of the climate.

Highest Temperature—

June.	July.	Aug.	Sept.	Oct.	Nov.	Dec.	Jan.	Feb.	Mar.	Apr.	May
137°	155°	152°	135°	124°	110°	112°	123°	114°	121°	124°	129°

Lowest Temperature—

June.	July.	Aug.	Sept.	Oct.	Nov.	Dec.	Jan.	Feb.	Mar.	Apr.	May.
110°	138°	133°	118°	104°	94°	97°	95°	101°	108°	98°	101°

We shall now consider our first class of climates, that of the snowy peaks of the Rockies, and the enterprise and energy of the American Government has supplied unequalled records in this respect, as for fourteen years they maintained the meteorological observatory at the summit of Pike's Peak, 14,147 feet above sea level, the highest observatory in the world, where regular and careful observations have been made by the officials of the Weather Bureau. Meteorological energy has been equalled by engineering, and a cogwheel railway, with a wonderfully adapted locomotive, conducts visitors to the summit during the summer, though the snow compels the closure of the line in winter.

Pike's Peak is of granite, and on the somewhat flat summit are strewn great blocks of this formation. Snow is always present in patches, but there are no glaciers. The barometer shows 17·54 inches of pressure.

A remarkable feature of the Pike's Peak records is the resemblance between the recurring annual phases of atmospheric pressure and of the air temperature.

The curves of these are not only alike in having a

single bend, but the maximum of both occurs in July, and the minimum of both in January.

The mean monthly pressure rises and falls ·016 inches for every degree Fahrenheit of the monthly mean temperature. Similar phenomena have been noted on Mount Washington (6,279 feet), another lofty observatory of the United States situated in New Hampshire, in the Eastern States. It has also been found that actual barometic pressure in the Rocky Mountains, generally at altitudes above 4,000 feet, attains its minimum in January and its maximum in July or August, and that the barometric phases are of the same kind, in reference to the annual mean, as the temperature phases at such stations. This phenomenon of atmospheric pressure is the reverse of that in parts of the United States at low elevation, and results, according to General Greely,[1] from the lower average temperature of the winter months contracting the great body of air, so that much of it is brought below the summit of the mountains, while in summer the reverse conditions obtain.

The highest temperature on Pike's Peak was 64°, noted on July 19th, 1879 : the lowest − 39°, noted December 21st, 1887 ; the mean temperature being 19·3°. This has ranged during the fourteen years from 17·9° to 21·9°.

It may be interesting to note that Mount Washington during the same fourteen years had a maximum of 74° and a minimum of − 50°, showing at a considerably lower altitude greater extremes.

[1] *American Weather.*

The maximum daily range of temperature on Pike's Peak takes place in July and August, being 14·3° and 14·2° for each month respectively. The minimum range is in December (11·6°). Thus the mean daily range on Pike's Peak is about half that noted on the prairie plains, on which Denver and Colorado Springs are situated.

The rainfall on Pike's Peak has been stated to be 29 inches, of which the minimum (1·39 inches) is in February, and the maximum (4·46 inches) in July. Thirty-three per cent. falls in spring, 35 per cent. in summer, the remaining 32 per cent. being distributed equally between autumn and winter, these seasons exhibiting the same dryness as at Denver and Colorado Springs.

The winds are chiefly westerly ; 31 per cent. are from the south-west, 20 per cent. from the west, 21 per cent. from the north-west, 10 per cent. from the north, 8 per cent. from the north-east. 5 per cent. from the south. and the rest are distributed among the east and south-east and calms.

This contrasts with Colorado Springs, nearly at the foot of Pike's Peak. where the most prevalent winds are the north (27 per cent.) and the south-east (22 per cent.) : showing the effects of local currents.

The velocity of the wind varies from a mean of 26·6 miles in January to one of 12·3 miles in August. The highest mean monthly velocity occurs with the lowest mean monthly temperature. and the least mean velocity with the highest mean temperature. Severe and prolonged wind storms are rare at the summit of Pike's

Peak, and the days are few when the mean hourly velocity equals or exceeds 50 miles. In two storms in fourteen years it ranged 70 and 71 miles an hour, and once, on May 11th, 1881, it reached 112 miles.

Mount Washington, on the other hand, has stronger winds and they prevail longer ; velocities of 116 miles have been recorded, and on one occasion the great rate of 186 miles.

The most interesting part of the meteorological diary on Pike's Peak is the record of the electrical phenomena, and as abridgment would spoil the pure American style in which the narrative is written, I give some extracts.

February 24, 1874.—Summit continued enveloped in snow clouds till 2 p.m. As a curious illustration of the rapidity with which water boils at this elevation, the observer notes that this evening a dish pan full of loose snow was set on the hot stove to melt, and in a very short time the water in the bottom of the pan began to boil, while the snow on the top of it was yet 3 or 4 inches deep. On examination it was found that a solid crust had formed above the boiling water, and even this was not sufficient to condense all the steam, which escaped with loud hisses from its icy confines.

May 29, 1874.—At noon the summit became densely enveloped and sleet fell in short showers till evening, while the rumbling of distant thunder was occasionally heard to the south. At 6 p.m. a storm began which will long be remembered by the present occupiers of the station. A violent thunderstorm, accompanied by heavy sleet, passed over the peak from south-south-east to north-north-west. It came up so suddenly that there was barely time to cut out the telegraphic instruments before blinding flashes of fire came into both rooms from the lightning arrester and stoves ; loud reports followed in rapid succession, while outside the noise of thunderclaps was almost drowned by the rattling of the indescribably heavy showers of sleet which followed each discharge. The storm appeared to surround the peak in all directions. At 7 p.m. a bolt struck close to the north window, and at the same time a heavy discharge took place through the lightning arrester, which made it appear that the building had been struck and was on fire. A cloud of smoke filled the embrasure of the north window, which was afterwards found to have resulted from the melting of rubber insulation on the office wires.

After this discharge the storm passed slowly to the west. The observers were beginning to breathe more freely when the wind veered to the south-west, and brought back the storm in all its fury. The rattling of torrents of sleet, mingled with the incessant rolls of thunder, the blinding flashes and loud reports in the rooms, were enough to make the stoutest heart quail.

It has been noticed, by closely watching the beginnings of storms which were so frequent during the latter part of May and not uncommon during June, that the great majority of them originated over the extensive parks west, south-west and north-west of this peak, and dividing it from the main range. The lower strata of air become powerfully heated, and are probably at this season of the year, when the surrounding mountains discharge their melting snows into the parks, heavily charged with moisture. The cold, heavy west winds descending the eastern slope of the main range, and wedging under the heated, moist, lower strata, might explain the frequency of local storms on the peak during the hottest part of the day.

February 14, 1881.—Fog in morning and afternoon. Four mock moons appeared on different sides of the moon at night, at equal distances from that body and from one another. These increased in intensity of colours until they were brighter than the moon itself, which appeared as if hidden behind a veil. By-and-by bows as brilliant as those of the sun appeared over each, which increased in size until they almost joined, and formed a perfect circle of the most brilliant colours. These gave way for double halos, the second being nearer to the first than that was to the moon. Both were of a beautiful violet tint. When the moon had risen about two-thirds of the distance between the horizon and zenith, these rings disappeared and a new one appeared much further removed from the body, the refraction of whose light produced these fantastic shapes. The new halo was much paler than those it had succeeded ; it was the precursor of one of the most magnificent refractions of light that any human being ever witnessed. Near the moon appeared two mock moons, shining like balls of fire, and in the horizon opposite these were reproduced, but in a milder colour. A halo of a mild pink colour passed through the plane of these four mock moons and intersected the main halo. To complete the display a bow appeared at the zenith, and this was as brilliant as a rainbow, and comprised all the colours of that bow of promise.

June 7, 1882.—During a violent thunderstorm, while sleet was falling a "singing" or "sizzing" noise on the wire was distinctly heard. At 8.45 p.m., on opening the door, the line on the summit was distinctly outlined in brilliant light, which was thrown out from the wire in beautiful scintillations. On near approach to the wire these little jets of flame could be plainly observed. They presented the appearance of little electrified brushes or inverted cones of light, or more properly, little funnels of light with their points to the line from which they issued in little streams about the size of a pencil lead, and of the brightest violet colour, while the cone of rays was of a brilliant rose white colour. These little cones of light pointed from the line in all directions, and were constantly jumping from point to point. There was no heat to the light, though it was impossible to touch one of these little flames, for as soon as they were approached by the finger they would instantly vanish, or jump to another point of the line. Passing along the line with finger extended, these little jets of flame were successively puffed out, to be instantly relighted in the rear. It was a curious and wonderful sight No sensation was experienced on applying the tongue to the line. Not only was the wire outlined in this manner, but every exposed metallic point or surface was similarly tipped or covered. The cups of the anemometer, which were revolving rapidly, appeared as one solid ring of fire, from which issued a loud rushing and hissing sound. The anemoscope represented a flaming arrow, and a small round wooden stake stuck up in the snow to show the position of the gauge was similarly tipped, as well as an angle of our stone chimney. Observer, on placing his hands close over the revolving cups of the anemometer, where the electrical excitement was abundant, did not discover the slightest sensation of heat, but his hands became instantly aflame. On raising them and spreading his fingers, each of them became tipped with one or more cones of light, nearly 3 inches in length. The flames issued from his fingers with a rushing noise. similar to that produced by blowing briskly against the end of the finger when placed lightly against the lips, accompanied by a crackling sound.

There was a feeling as of a current of vapour escaping. with a slight tingling sensation. The wristband of his woollen shirt, as soon as it became damp, formed a fiery ring around his arm, while his moustache was electrified so as to make a veritable lantern of his face. The phenomenon was preceded by lightning and thunder, and was accompanied by a dense driving snow, and disappeared suddenly at 8.55 p.m., with the cessation of snow.

June 9.—Repetition and previously defined sensations, and the observer's hair stood erect, crackled, and the pricking sensation to the scalp (bareheaded) was extremely painful. The peculiar electrical odour was strongly recognised. To protect his head, he put on his black felt hat and returned to the roof. But a few seconds elapsed before he was fairly lifted off his feet by the electrical fluid piercing through the top of his hat giving him such a sudden and fiery thrust that he nearly fell from the roof in the excitement. Instantly snatching the hat from his head, he observed a beam of light, as thick as a lead pencil, which seemed to pass through the hat, projecting to about an inch on either side, and which remained visible for several seconds. The top of his hat was at least two inches from his head when this fiery lance pierced him. When the fluid began to thrust its fiery tongues into other parts of his body, he was spurred to a hasty, but " brilliant " retreat. He experienced a peculiar burning or stinging sensation of the scalp for several hours afterwards. The phenomenon lasted fifteen minutes. Lightning and thunder were continuous during the evening ; the sharpest flashes drew from the metallic roof loud snapping sounds.

November 17, 1882.—Aurora this morning at 4.30 a.m. The arc was about 12° high in the centre. The one end extended nearly to the eastern horizon, the western one being hidden by the mountains. Along the eastern part luminous beams shot up from 15° to 20°. For a time it looked like a veil or sheet of pale light, but before fading it became very bright. A wide band of light from the sun extended nearly to the zenith, although nearly two hours before the sunrise. Ended 6 a.m. About 4.45 a.m. the "singing" or humming noise began on the telegraph wires, and became very loud. Aurora began to-night at 6.15 p.m. It appeared in the form of an irregular greenish-white cloud along the northern horizon, and under it was a very dark space. Above and apparently issuing from the green light, in different places, was an intensely red light, almost of a blood colour. These red spaces were probably 15° wide and from 15° to 18° high. Near these, beams of white light extended to the zenith.

The red would alternately fade and reappear, until finally it remained stationary at the west end of the clouds, fading and appearing in rapid succession until 8.30 p.m., at which time the white cloud appeared to break up into beams or groups of beams, which were probably from 15° to 20° in height. At this time the red light at the western part was intensely bright, and all gradually faded, the white colours settling down into a regular arch of about 15° high in the centre, and remained so. Aurora steadily disappeared at 11 p.m.

We next come to the climate of the natural parks, varying in altitude from 7,000 to 10,000 feet. Unfortunately I have only imperfect meteorological records, as there has hitherto been no station of the Weather Bureau in North or Middle Park. One has been recently established in South Park near Como, but I do not know its altitude, and the information from there, except as to rainfall, is incomplete. These parks extend from 37° to 41° north latitude and are of different elevations, so that considerable choice of climate may be found in them. In many places the vegetation shows an excellent soil, a good supply of water, shelter from winds, and freedom from the great cold of the higher peaks.

The Weather Bureau gives, among reports by voluntary observers, a year's observations made at San Luis, a small town situated at an elevation of 7,946 feet in San Luis Park, at the extreme south of Colorado state, and these I have analysed with the following results :—

The annual mean temperature is 41·7°, the mean maximum 70°, the temperature rising in June and July to above 90°, and falling in December and January to

— 25°; the mean minimum being 8·28°. The annual
rainfall is 13½ inches, occurring principally in September, December (when 2½ inches fall), March and April.

At Como (Middle Park) the rainfall is less, being
11·64 inches, but it occurs chiefly in summer, as on
the plains.

I explored Estes and Manitou Parks, and perhaps
an account of my visit to one of them may enable
you to picture them to yourselves better than a minute
description. Let us take Estes Park, which is close
to Long's Peak, and the greater part of which belongs
to an English company, of which Lord Dunraven and
Captain Whyte are the principal directors.

We left Denver one fine October morning, and after
three hours by rail, journeying northwards, we reached
Lyons, a small timber-built town in the foot-hills, the
present terminus of the railway, which it is contemplated to continue up to Estes Park itself. Quitting the
rail we ascended an excellent road in a winding valley,
through which the Little Thompson River flows out
from Estes Park. We first passed some grand masses
of red sandstone, which, as is common in the Rockies,
rise precipitously, ending in pinnacles resembling
castles and ramparts of fortresses and the like, and
soon we reached the rocks of granite formation.

The sinuous valley was well wooded with scrub oak,
in October of bronze to scarlet hue, with golden-tinted
cotton trees, a species of poplar very common in the
States, with what they call cedars, a sort of cypress ;
and strewn about were huge boulders, brilliant with
ferns and mosses, and wreathed with clematis, which

had fallen from the beetling cliffs on either side. As we
drove upwards the cotton trees and cedars gave place
to pines of various kinds, some of great size, and here
we made the acquaintance of some of the natives, the
gray squirrel and the chipmunk, a lively little animal
about half the size of our squirrel, with a long bushy
tail striped in two shades of grayish brown. They
were skipping about the larch fences in which they
delight. The beautiful American jay, with its brilliant
blue plumage, flew from tree to tree. We passed the
entrance to Antelope Park, another fine pasture land,
and came into more open country with numerous
ranches; before each farmhouse was a pile of stag-
horns shed by the deer in the adjacent mountains, and
collected by the farmers.

After four hours' ascent through exquisite scenery
we reached the portals of Estes Park, and found our-
selves descending into a magnificent basin of park-like
country, interpersed with several species of pines, and
backed by grand mountains.

The park is an irregular shaped depression with
various recesses, and measures ten miles in its greatest
diameter and four miles in its smallest. It is undulat-
ing and carpeted with excellent grass, but some fine
wooded eminences rise in parts. Surrounding it is a
remarkable circle of rocks ; behind lie the grand mountains : to the south Long's Peak (14,271 feet), to the
west Mount Upsilon, then Mount Kenry (named after
the Earl of Dunraven), and the Mummy Mount, and
on the north, Mount Signal, and to the east the
granite form of Mount Olympus.

The Thompson River, a moderate sized trout-stream, flows through Estes Park, and there are several small lakes. On the south side rises a fine wooded hill called Prospect Hill, up which we saw a coyote stealing (a coyote being an animal something between a wolf and a fox).

The park itself stands at an elevation of 7,500 feet, and contains an hotel and Captain Whyte's house—where we were most hospitably entertained—as well as several ranches, and in summer a large number of visitors camp out for shooting and fishing purposes. There is plenty of fish and game. Deer had been down in the park the day we arrived, and, what was more rare, a herd of mountain sheep had appeared on one of the crags above Captain Whyte's house. Beavers, of which we saw the traces, abounded, and had built their dams so effectually as to cause some trouble by diverting part of the stream destined for irrigation. Coyotes skulk about but do not seem to do much harm, though the wire meat-safes must always be placed at a height beyond their reach. We drove across the beautiful park in several directions, generally over the grass, then burnt up after the long summer, and we selected the site for a proposed new hotel, and saw Captain Whyte's fine herd of 400 red Herefordshire cattle, all in excellent condition. He told me that they fatten on the good herbage, and lie out all the winter without shed or stable. The cattle apparently were very healthy, and had become thoroughly acclimatised. That night we had a sharp frost, which covered the ranche's pond with $\frac{3}{4}$ inch of ice, but hardly whitened

the trees, and before sunrise next morning I witnessed
a splendid Alpine glow on the surrounding peaks, and
then the sun rose gloriously in a clear sky, flooding the
whole park with the golden beams, and imparting
plenty of warmth to us all. From the information I
could gather from the residents, the climate is never
excessively cold, as appears in the San Luis Park
observations, and there is occasionally strong wind, the
mountains, which do not overshadow, to a great extent
sheltering the park. The rainfall is small, and snow
does not lie on the ground for any time.

From some recent observations made in the park
by Dr. Carl Ruedi (late of Davos), and compared with
similar ones taken at the same time at Davos, it would
appear that the relative humidity percentage is much
less at Estes Park, that the rainfall is smaller, and the
number of rainy days fewer, and that the hours of
sunshine during the winter months are considerably
longer than at Davos.

A great advantage are the endless excursions, not
only in the park itself, but into the beautiful valleys
which open into it, the Black Cañon and the Horse
Shoe Valley for instance;—excursions replete with
interest for the artist, the sportsman, and the man of
science.

At present these parks are used chiefly in summer,
when the dwellers in cities, like Denver, flock to them
and camp out during the two hottest months, but it is
contemplated to utilise them during a longer part of
the year, as being well suited for more active and less
delicate invalids. There is at present accommodation

in Estes Park, but more is forthcoming in the shape of
a new well-appointed hotel to be kept open all the year
round. The present hotel has 50 beds and several
ranches with cottages attached: Ferguson's, Macgregor's
and James's (late Elkhorn) also lodge invalids with
tolerable comfort and at a moderate cost. There is a
post daily bringing letters and parcels.

Manitou Park lies 8,000 feet high, 26 miles from
Colorado Springs, and rather nearer to Manitou, up
the Ute Pass, with a fine view of Pike's Peak. It is
about 10 miles in length and 4 in breadth, and is one
of the approaches to Middle Park.

The scenery much resembles that of Estes Park, and
the neighbourhood abounds in large game, such as elk
and deer. There is an excellent hotel and a number of
wooden cottages, which are used by visitors in summer,
when the place is a great camping centre.

A picturesque line of rail, the Colorado Midland,
connects Woodland Park Station with Colorado
Springs, and then a 9-mile drive, partly through
lovely fir woods, leads to Manitou Park, which is
much recommended by the medical men of Colorado
Springs.

Our third class of elevations, the foot-hills, an
expressive term, may be considered in conjunction with
the fourth, the prairie plains, the difference being that
the climate of the former is rather cooler.

The third and fourth class comprise all the Colorado
plains of an elevation of 5,000 feet and upwards, and
the towns situated on them, such as Denver, Longmont,

Boulder, Golden, Colorado Springs, Manitou, and the various settlements on the foot-hills.

The climates of Denver and of Colorado Springs, as set forth in the admirable meteorological reports of the Weather Bureau, have been well examined by Drs. Denison and Solly, and may be considered as typical of the true Colorado climate, the principal features of which have been already given.

The prairie in which these towns lie is as remarkable for its vastness as for its colouring. When standing on any elevated spot, one sees it is not a flat surface, but presents here and there undulations, though stretching away, as it does, for hundreds of miles north, east and south, these are sometimes hardly visible, and the general impression is of some great billowy moving sea which has suddenly become petrified.

The sun rising in the east floods the vast plain with its rays and turns it a brilliant golden colour beautiful to behold, and gradually as the shadows extend, tints of red, purple, and brown appear and advance in great lines across the plain, while with the setting sun the colouring becomes first orange, then red, then purple, changing at last from delicate pearly tints to cold gray. The sky also undergoes the most brilliant changes, passing from the usual sunset phases to an exquisite violet which suffuses successive portions of the sky long after the sun has set. The vault of heaven appears of boundless height, and the air is so transparent that objects twenty miles off appear close at hand, and Pike's Peak, 75 miles distant from Denver, is quite distinct; indeed some of the fine peaks seen

from that city are calculated to be more than 120
miles off.

Another feature of this vast plain is the absence of
life. You may travel for miles and miles and see
nothing but an occasional prairie dog village, with its
little fat denizens sitting up on the tops of their mounds
with their paws hanging in front, looking like posts, so
unmoved are they, even when the locomotive rushes
past ; after a while a few cattle are to be discerned, then
a mounted cowboy, looking larger than human from his
solitariness—but the general feature of the prairie is its
intense solitude.

The vegetation consists of buffalo grass, several
varieties of cactus, a small kind of yucca, and what is
called the prairie flower, besides lilies and other summer
blooming flowers.

Driving in a light buggy with a pair of fast cobs is
very pleasant, provided you hold on tight, as it is not
uncommon to descend into a prairie dog's hole, or
worse still into a dry creek or water bed, where the jolt-
ing is considerable, for the horses delight in the grassy
trail, and fly through the air at a tremendous pace.

As I have said before, the prairie is undergoing
cultivation ; by means of sinking artesian wells and
bringing water from the mountains, the plains are
gradually being supplied with irrigation water, and are
made to produce a better grass as well as wheat and
Indian corn ; a little less cactus, a little more alfalfa
(lucerne) everywhere, this last being the usual western
food for cattle ; but it will be long before the prairie is
transformed into great farms.

Denver is situated in the prairie on the small river of South Platte, and though only about thirty years old, is a city of 150,000 inhabitants with fine buildings, capital schools, excellent clubs, theatres, monster hotels, cable and electric cars, electric lighting, and all the advantages of American civilisation. The streets are well planned and many are paved with asphalte the avenues are wide and frequently lined with trees and command fine views of the Rocky Mountains. The city extends over at least five miles, and though parts of it are smoky, owing to the ore smelting and other works, the suburbs, which are remarkably open, and only bounded by the prairie, are charming and suitable for invalids' residence. The medical profession is well represented, while a completely equipped faculty of medicine exists in the University of Denver, and I may add that many of these doctors are instances of consumption cured by the climate.

There are also a number of boarding houses in the small towns and ranches in the foot-hills and up the cañons, where Denver medical men place their patients with advantage, such as the following :—Stewart's ranch in Bear Cañon (7.000 feet), about 20 miles from Denver, Longmont (5.000 feet), a small town with good water and lighting, and near it an excellent moderate *pension* called Hygieia. Boulder (5.400 feet), at the mouth of the romantic Boulder Cañon, is a small town with a university, 29 miles from Denver, and has suitable accommodation, with charming excursions up the adjoining valley. Greely and Fort Collins, at a little lower elevation and some 50 to 60 miles to the north of Denver, have been well spoken of for invalids.

Idaho Springs (7,800 feet), in the Clear Creek Cañon, is strongly recommended by Dr. Denison, and stands on a plateau well sheltered by mountains, having remarkable saline and ferruginous springs, used for baths and drinking, and, like most of the preceding, is connected with Denver by rail. There is a very good hotel at Perry Park (6,000 feet), situated on a spur of the Rockies called the Divide, which separates the basins of the South Platte and the Arkansas rivers, about 40 miles south of Denver, and there are many others, one of the great advantages being the extensive rail communication in all directions with Denver.

Colorado Springs is situated on the prairie, at an altitude of 6,022 feet, 75 miles south of Denver, 5 miles from the foot-hills of the Rockies, and 6 miles from the base of Pike's Peak, which forms so fine a feature in the view from the town. It has a population of 13,000, and no manufactories, and consequently no smoke. The town is laid out picturesquely in avenues from 60 to 120 feet in breadth, lined with a double row of trees, which run north, south, east, and west, thus intersecting each other, the main roads being traversed by electric cars. The buildings are handsome, and the private houses are specially artistic, many of red sandstone and surrounded by gardens. There is a top soil of 2 feet and gravel and sand to an average depth of 60 feet below, and consequently all moisture drains away rapidly. The main drainage and the plumbing—as our American cousins call it—is good, and the water supply is excellent, the water being brought from the mountains in iron pipes.

To the north of the town lies the Divide, which gives some protection from northerly winds, to the south the open prairie, to the east the prairie on which rises a low range of hills called Austin's Bluffs, and to the west and south-west the great masses of Pike's Peak, Mount Rosa—named after Miss Kingsley—and the beautiful Cheyenne Mountain, with the foot-hills in front of them.

Intervening between these grand mountains and the town are Colorado City, and rather to the north of it the celebrated Garden of the Gods, Monument Park, and Glen Eyrie, with the fantastically shaped red sandstone rocks ; while nestling under the very mountains, surrounded by more timber than is usual in Colorado, lies Manitou Springs, about 1,000 feet higher than Colorado Springs, where the mineral springs really do exist and are of delicious quality, whereas there are none at Colorado Springs, the title of which is a misnomer.

The mean temperature of the Colorado Springs is 46·4°, but it is composed of considerable extremes. The maximum rises to 101°, the minimum falls to −25°, but curiously the extremes do not appear to be much felt, perhaps owing to the dryness. The mean rainfall for 10 years is 15·87 inches, and of this about 12 inches fall in spring and summer, generally in thunderstorms ; only 4 inches are noted between September and March. Snow, as a rule, disappears by evaporation in the sunshine. The number of clear days is 194, of fair days 128, and of cloudy 43. The sun shines during 330 days of the year, and for 165 days out of the 182 from October 1st to April 1st, so that on an average an invalid is deprived of sunshine for less than half-a-day a week in winter. The power of the sun is great, and during the entire winter ladies need parasols and invalids sit in the open piazzas, which are a feature of the houses, without extra wraps. The wind is the most troublesome item of the climate, and generally rises in the afternoon. The annual mean velocity is 8·58 miles per hour.

Dr. Solly[1] gives an admirable account of an invalid's day in midwinter, *i.e.*, from 9 a.m. to 4 p.m., which I venture to quote.

"After a night in which there has been a hard frost and a clear sky, with a light breeze from the north, and during which the invalid has usually slept soundly under several blankets, with his window partly open, he wakes up to find the sun shining at his eastern window. And this is a feature which, whatever the weather may be later in the day, is rarely absent. After breakfast our invalid steps into the street, being then in an atmosphere in which the heat in the sun is 92°, and in the shade 30°. A gentle air is stirring from the north-east at the rate of 8 miles an hour. The mean dew point is 18°.

"As the day proceeds the temperature rises to its highest point, between 3 and 4 p.m. being 100° in the sun, and 40° in the shade, while the wind, which has veered rapidly from the north to the south, blows with its highest daily velocity of 13 miles an hour. After 2 p.m. the wind works back again towards the east, being at sundown north-east, and continuing as darkness falls to shift back to the northern quarter, whence it blows from 8 p.m. to 9 a.m., its velocity dropping to between 7 and 8 miles an hour: the temperature of the air at the same time falling 3° to 4°.

The ground is usually bare of snow; no rain falls from mid-September to mid-April, and the sun shines unobstructed by clouds.

[1] *Facts, Medical and General, about Colorado Springs.*

During the three winter months the cloudy days do not average more than three a month.

Accommodation is excellent and plentiful, and there are several able and experienced medical men, including Dr. Solly, to whom Colorado Springs owes so much. The excursions to be made are endless and charming. The cañons of the Cheyenne Mountain present lovely scenery of trees, granite needles, and waterfalls, the Ute Pass is grand, but the finest of all is Pike's Peak itself, with its neighbouring valleys, which is daily ascended by rail in summer. Riding and driving are the chief exercise used, and the Broadmoor Casino, with its boating on the lake, its music, and its races, furnishes amusement, while for the vigorous, the mountains and parks within easy distance offer plenty of shooting and fishing.

Manitou Springs is somewhat better protected by the mountains from wind, though it enjoys sunshine for a shorter period of the day than Colorado Springs. The annual mean temperature is 47·3°, the maximum, 96°, occurring in July, and the minimum, 23°, in January. The relative humidity is 54 per cent. The rainfall, of which I was unable to procure statistics, is probably low.

It is a pretty little place, with some most valuable mineral waters and excellent hotels. It is also the starting point for several suitable summer stations on the Ute Pass, such as Cascade Cañon, Ute Park, Green Mountain Falls, and Woodland Park, all of which have good hotels placed at various elevations above Manitou.

The above description will, I hope, give a sample of the climate of Colorado and afford some explanation of its probable factors.

The chief elements appear to be—

1. Diminished barometric pressure, owing to altitude, which throughout the greater part of the State does not fall below 5,000 feet.

2. Great atmospheric dryness, especially in winter and autumn, as shown by the small rainfall and the low percentage of relative humidity.

3. Clearness of atmosphere and absence of fog or cloud.

4. Abundant sunshine all the year round, but especially in winter and autumn.

5. Marked diathermancy of the atmosphere, or, as Dr. Denison expresses it, " the increased facility by which the solar rays are transmitted through an attenuated air," producing an increase in the difference of sun and shade temperatures varying with the elevation in the proportion of 1° for every rise of 235 ft.

6. Considerable air movement, even in the middle of summer, which promotes evaporation and tempers the solar heat.

7. The presence of a large amount of atmospheric electricity.

Thus the climate of Colorado is dry and sunny, with bracing and energising qualities, permitting outdoor exercise every day all the year round, the favourable results of which may be seen in the large number of former consumptives whom it has rescued from the life of invalidism and converted into healthy active workers. Its stimulating and exhilarating influence may also be traced in the wonderful enterprise and unceasing labour which the Colorado people have shown in developing the riches, agricultural and mineral, of their country. Let us take the latter alone : In 1890 30 millions of dollars, or £6,000,000 sterling, worth of precious metals was mined in the State of Colorado, which is larger than the United Kingdom and Ireland and at present numbers only 500.000 population. Thirty years ago Denver may be said not to have existed. Now (1894) it is a well-built, well-organised city of 150,000 inhabitants.

So much for Colorado, and had time sufficed I would have visited some of the other high-lying sanitaria of America, such as those of New Mexico, or again the more moderate ones of the Adirondacks in the Eastern States, which have been commended by Dr. Loomis of New York, where, under Dr. Trudeau's and his direction, an admirable sanitarium has been erected, and the treatment of consumption carried out systematically and successfully.

INDEX.